silver surfers'

COLOUR GUIDE TO

online.
shopping

roger shaw

foulsham

foulsham

The Publishing House, Bennetts Close, Cippenham, Slough,
Berkshire, SL1 5AP, England

Foulsham books can be found in all good bookshops or direct from
www.foulsham.com

ISBN: 978-0-572-03310-1

Copyright © 2007 Roger Shaw

Cover photograph © Superstock

A CIP record for this book is available from the British Library

The moral right of the author has been asserted

Printed in Dubai

Contents

Introduction	7
Shopping on the internet	8
Can I really save money by shopping on the internet?	9
What is an internet shop like?	9
Internet auction sites	10
This book has been written especially for you	10

Chapter 1 Web Accessibility	13
Accessibility options	13
Resizing text when surfing the internet	14
Further help with accessibility	15
Useful websites	16

Chapter 2 Protecting Your Computer	17
Preventing viruses in e-mails	17
Using a firewall	18
Anti-virus software	22
Buying firewall or anti-virus software	22
Protecting important documents and pictures	24
Useful websites	25

Chapter 3 Security on the Internet	26
Fake e-mails	26
User names	29
Passwords	29
Secret question	29

Chapter 4 How to Search for Internet Shops	31
Search engines	31
Using a search engine	32
Choosing the right keywords	33
Refining your search	34
Web-friendly sites	35

Chapter 5 Price Comparison Websites   37
Using price comparison websites   37
Comparing prices   38
Useful websites   42

Chapter 6 Paying for Items from the Internet   44
Debit cards   44
Credit cards   44
Personal cheque   45
PayPal   45
Using debit and credit cards   46

Chapter 7 Using PayPal   48
Registering with PayPal   48
Your PayPal account   52
The verification process   53
Adding funds   59
Paying through PayPal   61
Sending money   62
Resolving payment disputes   63

Chapter 8 Shopping for Groceries   65
Registering on a supermarket website   66
What to expect from a supermarket website   70
What you will find on the groceries page   71
How to shop for groceries   77
Paying for your shopping   80
More grocery shopping websites   82

Chapter 9 Household Goods and Services   85
Shopping for electrical goods   85
Washing machines   86
Dishwashers   88
Fridge freezers   89
Televisions   89
Computers   90
Gardening   92
Books   94
Gas and electricity   95
Insurance services   97
Useful websites   98

Chapter 10 Shopping for Clothes   **104**
Types of account   104
Size guides   104
Returns   105
Ordering clothes online   106
Buying clothes on eBay   109
Useful websites   110

Chapter 11 Health and Mobility Products   **112**
Keeping active   112
Exercise equipment   113
Mobility aids   113
Buying a mobility scooter   114
Health products   118
Private health care   118
Useful websites   119

Chapter 12 Holidays   **123**
Should I book an internet holiday?   123
Package holidays   124
Non-package holidays   124
Self-drive holidays   125
Booking a trip through the Channel tunnel   126
Booking flights   127
Airport parking   128
Useful websites   129

Chapter 13 Auction Sites   **131**
eBay auction site   131
How eBay works   132
Getting started   133
Becoming an eBay member   133
Choosing your user ID   135
Your eBay password   136
Finalising your registration   138
Signing in and out   141
My eBay   143
Searching eBay   147
eBay payments   152
Resolving payment disputes   153

Protecting your account against fraud — 154
Useful websites — 157

Chapter 14 Buying On eBay — **158**
The bidding sequence — 158
The path to success — 159
The winner's curse — 161
What not to buy — 161
Who not to buy from — 162
Checking feedback — 163
Choosing your item — 165
Knowing the seller — 165
The reserve price — 167
Checking postage and packing — 168
Watching items — 168
Bid history — 169
Comparing auctions — 170
Buy it now — 172
Best offer — 173
When to make a bid — 174
Sniping for beginners — 174
Making a bid — 175
Bid confirmation — 178
Checking your bid throughout the auction — 179
Bid retraction — 179
Item not won — 182
How to pay — 183
Leaving feedback — 184
Resolving problems — 185

Index — **188**

Introduction

We live in a world in which we are surrounded by technology, often with its own unique language, a language that is always growing and that changes so quickly that it sometimes seems a hopeless task to keep up with it all.

Consider that only a few years ago:

- A mouse was a little animal;
- Windows were something you looked through;
- A keyboard was something you played to make music;
- A 3.5in floppy was something men didn't talk about;
- A hard drive was from the trip from London to Edinburgh;
- Spam was a kind of tinned meat.

You will be pleased to know that within this *Silver Surfer* series unnecessary jargon has been avoided. Forget what you may have heard about how complicated computers or internet shopping can be. You will soon be able to achieve what you want to achieve, because this book will show you how.

I often hear people say, 'I don't understand computers' or 'I'll never use one.' But in reality, all of us come into contact with computers every day because just about every product – from a toaster to a car – is controlled to some extent by the same kind of microprocessors that you find in computers. The television in your home contains a computer, together with some very complex electronics. But you don't need to understand how those electronics work in order to enjoy your favourite programmes.

The same goes for your computer and its features. Modern computers are not designed for people who want to know how the background programming works. They are made to be easy to use by people like you who just want to make life a little easier. So don't be nervous about embracing new technology. It's all quite simple: trust me!

Nor should you be intimidated by youngsters who seem to know everything without reading the instructions. Apart from having the lightning-quick reactions of youth, they have the advantage of being brought up with new technology, so they simply take it for granted and are not afraid to try things out and make mistakes. You might take a slightly more cautious approach, but you will be pleased to know that it is our generation that is set to dominate the internet, and in particular online shopping. Take your time, follow the instructions and you will soon wonder what you ever worried about in the first place.

Shopping on the internet

For thousands of years, our ancestors haggled and traded to obtain the lowest prices. Today our opportunities to haggle over prices are fairly limited and mainly confined to open-air markets and car boot fairs. Today we shop by going from one store to another, comparing the various prices on offer. The retailers do the same to make sure their prices are competitive, so the chances of making a significant saving are pretty slim. Shopping on foot locally also restricts your choice.

There is another way: online or internet shopping. Using your computer and an internet connection, you can browse through thousands of shops anywhere in the world. You can compare prices instantly and have your chosen goods delivered to your door. Just about everything is available on the internet now, from your weekly shop to luxury goods.

Can I really save money by shopping on the internet?

The answer is a huge yes! The saving could well be huge, too, compared with high street prices.

How can internet shops offer such huge savings?

Quite simply, online shops do not have the large overheads that are associated with a high street store, so they can pass these savings on to you, the consumer.

Consider what your high street store has to pay:

- High rent: the better the location, the more expensive the rent and rates will be;
- Staff wages: including both shop-floor staff and management;
- Shop fittings: to make sure the store looks attractive enough to draw people in;
- Heating and air-conditioning: to make the shop comfortable to shop in;
- Advertising and promotion: so you know where the shop is and will call in to check out what it has on offer.

The cost of all this is passed on to you, the customer.

What is an internet shop like?

Typically, many online stores are huge automated warehouses where your order is picked and despatched efficiently.

At the other end of the scale, specialist internet shops can be operated from home and run on a cottage-industry basis, allowing you to access goods or services that it would otherwise be impossible to bring to market economically.

Internet auction sites

If you enjoy looking for a bargain combined with the excitement of a real auction, then an internet auction site is something you should try. There are many available on the internet, although eBay is the biggest and the one most people have heard of.

This book has been especially written for you

I have worked with many older students and always find them practical, enthusiastic and eager to learn. I cannot understand why so many so-called 'guides for older people' only tell half the story and dumb down the information. That is neither helpful nor sensible. The only difference I find between my young and my mature students is that my mature students are sometimes just a little slower and need just a little more reassurance and practice, whereas my young students seem to rush in, skip many key points and do not fully understand what they are doing. I think it would be true to say that my mature students, in the main, achieve a greater depth of knowledge than my young ones!

Within this colour, step-by-step guide I have presented all the information you need on security, internet shopping and auctions in bite-size pieces, allowing you to learn at your own pace and to understand fully what you are doing. I have packed the book with lots of real examples for you to try for yourself, together with extra tips and 'Don't worry' boxes. These will reassure you about things or explain extra bits that you might come across but that are not essential to understand at that stage. If you are comfortable with the section you are learning, then just ignore them. However, if something is troubling you, reading them might just provide you with the answer.

Within each chapter you will find a reminder box of the key points covered. This allows you to identify quickly topics that you have not understood or have forgotten, giving you the opportunity to revisit them before moving on to the next chapter.

In addition to this, the important issues of security and finance are covered. Before you start shopping and spending your money it is really important to protect yourself from possible misuse of your financial information. Just as you guard your purse or wallet, any account details or passwords should be similarly protected, and this book includes simple ways to achieve that.

Computer terms

Here are a few terms you will find useful when you are looking through the book. Some of them will be familiar, but you may wish to refresh your memory on others.

Accessibility Options that you can change in Windows to help you see your computer's information more clearly.

Advanced search In-depth search for specific items.

Bidder An eBay member who takes part in an auction.

Browser Also known as a web browser, this is simply the program used to get on to the internet. The most common browser is Internet Explorer.

Checkout The page on an internet site that displays your purchases, together with the total cost and a method of payment.

Firewall Software that helps to prevent access to your computer by other people whilst you are connected to the internet.

Hardware All the parts of the computer you can see.

Hyperlink A link from one page on a website that, if you click it, takes you directly to another page. It is often simply called a link.

Icon The picture that indicates what a link does if you click it.

Menu A list. The 'main menu' appears across the top of the screen. If you click one item, another menu will drop down.

PayPal An electronic method of payment using your credit or debit card or your bank account.

Screen reader A software application that attempts to identify and interpret what is being displayed on the screen for visually impaired users.

Shopping basket An imaginary basket used on websites to help keep track of your purchases before you visit the checkout.

Software Programs on a computer.

Spoof e-mails E-mail requests sent by criminals asking you to reveal your account details.

Spyware Programs that monitor your internet activity without your knowledge.

USB memory stick A plug-in device for holding data and pictures.

Virus A computer program designed to cause damage to information on your computer.

Chapter 1

Web Accessibility

Many of us struggle to read small print. I have a business associate who cannot read a thing without his glasses and he has a habit of forgetting to bring them whenever we are out at lunch, so I enjoy reading the menu and making up dishes I know he would not like! But seriously, if you find it difficult to read your computer screen there are some simple adjustments you can make yourself that will enlarge the text and enable features that will make your screen appear much clearer and easier to read.

Accessibility options

Making these changes only takes a few minutes but could make the internet much more enjoyable and accessible.

- With your computer switched on, simply click **Start** in the bottom left-hand corner of your desktop.

- Click **Control panel** and this window will open up.

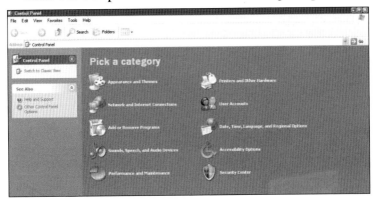

- Click **Accessibility options** and you'll see this window.

- Now click **Configure windows to work with your vision, hearing and mobility needs**.

The program will ask you a series of questions in order to best configure your computer to your individual needs, depending on whether you want to make adjustments to the display or perhaps to the keyboard.

Resizing text when surfing the internet

While you are browsing the internet, you can alter the size of the text on any web pages you are looking at. You can do this through your browser, which for most people will be a program called Internet Explorer.

- On the main menu across the top, click **View**.

- On the menu that drops down, click **Text size**.

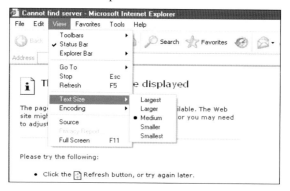

• Select the size that best suits you and click on it.

If you do not like the result or you change your mind, simply go back and do it again.

If you are using a different browser – for example, Netscape, Mozilla or Firefox – you use a slightly different process.

• From the main menu, click **View**.

• Click **Increase text size**.

• To decrease the size of the text, select **Decrease text size**.

Don't worry **about keyboard shortcuts**

There is usually more than one way of making things happen on a computer, and you can often activate options by using the keyboard instead of the mouse. To increase the text size, you can press the **Control** key, then press the + key. Repeat until you can easily read the text. To decrease the size, press the **Control** key, then the – key.

Further help with accessibility

If you are visually impaired, deaf or have mobility problems, there is plenty of help out there for you. A solution to your access problems could be just a click away. Many software companies will allow you to download a free trial of their products so you can try them and really make sure they match your needs before you think about purchasing. For example, a screen reader is a software application that attempts to identify and interpret what is being displayed on the screen. This is then presented to a blind user as speech, in what is known as a text-to-speech program or by driving a braille display.

Don't worry **about accessibility**

If you have trouble reading your computer screen, there are plenty of programs available that might help you. See the useful websites on the next page.

Useful websites

- **www.abilitynet.co.uk** National charity helping disabled people to use computers.

- **www.ageconcern.org.uk** The UK's largest charity working for older people.

- **www.freedomscientific.com** Connect Outloud 3 is a software package for accessing the internet with speech and Braille output. Jaws for Windows is a screen reader. Magic is software that magnifies the text on the screen to make it easier to read.

- **www.gwmicro.com** WindowEyes is a screen reader program.

- **www.rnib.co.uk** Royal National Institute of the Blind.

- **www.rnid.org.uk** Royal National Institute for Deaf People.

- **www.webbie.org.uk** A free, specialised text browser developed by the University of Manchester.

Your reminder box

- If you need help seeing your screen, first check your settings in **Control panel**.

- If you need to purchase a program to help you, then go for a free trial – this way you can make sure it's right for you before you buy.

Chapter 2

Protecting Your Computer

There is a lot of talk in the media about computer viruses and this is a subject of great concern to internet users. However, there are many things you can do to maintain the security of your system and protect yourself from infection.

If you understand the basics, it will help you to take the necessary action. Once you connect to the internet, your computer is at risk from viruses and spyware, so it is important to protect yourself. Viruses are programs designed to corrupt data – the information on your computer – and cause your computer to fail in one way or another. Some viruses contain spyware, which can be used to gain access to your financial or account information. Programmed to hunt out passwords, user names and to monitor keystrokes for unusual combinations of letters and numbers, they immediately send such information back to whoever planted the spyware. This can all be done without your knowledge and with no symptoms appearing on your computer.

Preventing viruses in e-mails

Your e-mail program, such as Outlook Express, will automatically warn you if it thinks an e-mail is suspect for any reason. You can also help prevent viruses sent via e-mail by only opening e-mails you are expecting or from people you know. Delete anything you are not sure about without opening it; if it is important, the sender will contact you again.

Outlook will put suspect e-mail in the 'Junk mail' folder instead of your 'Inbox'. Open this folder and look at the names without opening the e-mails. If they *are* junk, just click once on them, then click **Delete**. They will go into your 'Deleted items' folder, from which you can permanently delete them.

Additionally, contact your internet service provider (ISP) and ask what help it can offer in protecting you from viruses.

Using a firewall

Firewall is the name given to hardware or software designed to prevent access to your computer by other people who are using the internet.

In addition to the commonsense precautions above, you should make sure your computer has a firewall and anti-virus software installed, and you should keep them up to date. Before you buy any additional equipment or software programs, check what comes as standard with your computer, because you might already be protected.

Hardware firewalls

Hardware is the term used to describe the physical components of your computer, and so there may be a firewall in the broadband router – the little box – that connects you to the internet. If you have several computers in your home that share an internet connection, the firewall in the router would protect them all.

If you have a firewall in your router, it will come with a CD and simple, onscreen instructions so you can set it up to do what you want. Simply put the CD into the drive on your computer and it will lead you through the necessary steps.

If you find this difficult, don't worry; you can use a software firewall instead. In fact, I think they are more effective.

Software firewalls

Software is another name for a computer program, which may already be installed on your computer, or you may have to install it via a CD or by downloading it from the internet.

If you have a relatively new computer, it will probably be running Windows XP Service Pack 2 – which is the basic program that operates the computer – and this includes a good firewall, so you will not need anything else. You can check if your computer is running this version of Windows.

• Click **Start** at the bottom right-hand corner of the desktop.

• Click **Control panel**.

• Double-click **System**.

• When the system control panel appears, under the word 'System' it will tell you what operating system your computer is using.

Switching on your firewall

You should check that your firewall is switched on and working.

- Click **Start** at the bottom right-hand corner of the desktop.
- Click **Control panel**.
- Double-click **Security centre**.
- Click **Windows firewall**.

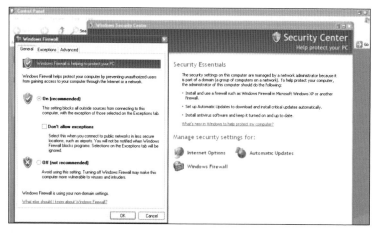

- The word 'On' should appear next to the green shield. If not, click the circle to switch it on.

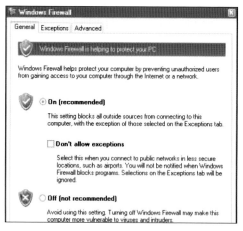

Keeping your firewall up to date

Because viruses are constantly becoming more sophisticated, you need to make sure you keep your virus software up to date. You can do this by going to the Microsoft website and downloading any updates.

- Open Internet Explorer, type **www.microsoft.com** into the address bar and click **Go**.

- From the menu on the left-hand side, click **Microsoft update**.

- Click **Start now**.

Microsoft will check your computer and offer you any updates. Simply follow the online prompts.

Commercial firewalls

If you have an older computer, it may be running Windows 95, 98, ME, 2000 or another operating system. If so, a commercial firewall is the preferred option. Commercial firewalls operate in the same way as a Windows firewall but generally give you extra control, information about how to configure the program and more support.

Security checks

If you are not sure if you need a firewall, you can check whether you have one installed with your operating system and that it is working by visiting Symantec for a free security check.

- Open Internet Explorer, type **www.symantec.com/home_homeoffice** into the address bar and click **Go**.

- Click **Free scan for virus**.

You can download a free basic firewall called ZoneAlarm.

- Open Internet Explorer, type **www.zonelabs.com** into the address bar and click Go.

- Click **Free online spyware detector**.

- Click **Click here to start your scan now**.

Anti-virus software

The second step to protecting your computer is to install anti-virus software. Without anti-virus software your computer is vulnerable to attack. Being infected by a virus can have serious consequences including fraud, loss of data, identity theft or a slow or unusable computer. It is therefore important to check that you are protected.

No anti-virus software offers total protection, but it should protect you from the main lines of attack by:

- Scanning incoming e-mails for attached viruses;

- Monitoring files as they are opened or created to make sure they are not infected;

- Performing regular scans of every file on the computer.

If you are looking for anti-virus software, there are a number of free programs available for personal use. In most cases, these free products are scaled-back versions of commercial products to which the software manufacturer hopes you will, one day, upgrade. You may wish to consider a trial version of software. The software company allows you between three and twelve months to use its product before asking if you wish to purchase the full item.

Buying firewall or anti-virus software

You can purchase a firewall or an anti-virus program individually or as a package combined with other protection software. When purchased as a package, this is sometimes called a security suite. A suite is cheaper than buying each individual program separately and tends to be easier to use. As with most things in life you get what you pay for, so be prepared to pay between £20 and £50 for a good security software package.

If you are going to purchase, then look at software from leading suppliers like F-Secure, McAfee and Symantec. Microsoft publishes a list of Windows-compatible programs. Their website adddresses are on page 25.

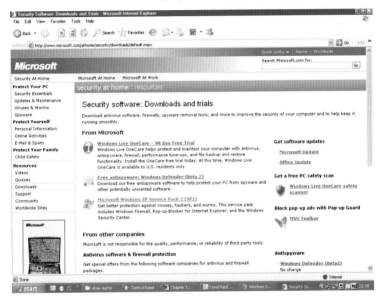

Software suppliers include high street retailers as well as online stores. I tend to download the program over the internet direct from the manufacturer's website, using my credit card to purchase. It is quick and easy, and by the time you have looked through the purchasing information in this book, you will feel confident enough to try it.

Protecting important documents and pictures

As with anything mechanical or electronic, things can go wrong with computers from time to time. It is usually relatively easy to resolve the problem, but just to be sure, you should make regular back-up copies of important documents and pictures on a CD or USB memory stick.

To save data on to a CD:

- Insert a blank CD into the CD drive of your computer. Disks are labelled either CD-R (for recordable), which means they can be used once only, or CD-RW (for rewritable), which means they can be used again.

- Open the window in which the name of the document you wish to back up appears.

- From **My computer**, open the CD-RW drive or DVD-RW drive (D:).

- You will now have two windows open. Click on the document, and whilst holding the mouse, drag the document to the CD window – you will see the icon move. Once in the CD window, release the mouse to 'drop' the file – you have now created a copy of this document.

- Click **Write these files to CD**. A CD wizard will open and prompt you to name the CD. It will then copy the files on to the CD and eject the disk when it is ready.

The instructions may vary slightly between computers. Remember to write the name of the files on the CD in indelible pen or on the CD case.

 Don't worry **about USB sticks**

USB sticks are devices for storing data in a way that makes it easy to back it up or to transfer it from one computer to another. They are tiny, efficient, very convenient and not expensive. They vary in capacity from 64 MB to 2 GB, so they can hold a huge amount of information.

You use exactly the same method to copy to a USB stick as to a CD, although you navigate to the removable disk (E:). Some computers may recognise the USB stick as removable disk (F:).

Don't worry **about firewalls**

If all this information about firewalls and anti-virus software has gone over your head a little, don't worry. It does not stop you from shopping on the internet. If you have a new computer, you will probably already be protected. If not, allow yourself time to get a little more familiar with things by working through the book, then come back to this section.

Useful websites

All these sites offer computer protection and security advice and downloads.

- www.free-av.com

- www.grisoft.com

- www.microsoft.com/athome/security/downloads/

- www.symantec.com/home_homeoffice/

- www.zonelabs.com

Your reminder box

- To protect your computer, use a good firewall and install anti-virus software.

- Keep them both up to date by visiting the manufacturer's website regularly and downloading any updates.

- Make back-up copies of important documents on a CD or a USB memory stick.

Chapter 3

Security on the Internet

The vast majority of people and businesses on the internet are honest. Sadly, however, as within any community, there are those who are less than honest. They, too, have internet access and might use some very clever spyware software to infect your computer to glean sensitive information. Although you should not be complacent about this type of crime, it is extremely unlikely to affect you if you have installed a good firewall and anti-virus software.

Believe it or not, most internet fraud occurs as a result of the user revealing his or her details voluntarily to the fraudster! There is no need for this to put you off, because if you follow some commonsense rules, the internet can be a very safe and secure place to shop.

Fake e-mails

I am sure that if someone knocked on your door and asked for your bank details, including your passwords and PIN, you would send them on their way empty handed. All you need to remember is to use the same judgements of common sense when using your computer as you would with someone who knocked at your door. Remember that although many computer systems are automatic, someone is behind the computer inputting and extracting data. Many frauds occur when criminals simply ask you to confirm your information in an e-mail, pretending to be from a bank or internet shop.

Within the e-mail they will use a copy of the organisation's official logo and even the reply address will appear genuine. All these fake e-mails have one thing in common: they ask for your password or bank details. To encourage you to reveal this, most e-mails will request you verify your account details as a matter of urgency to prevent your account being restricted or closed. Some will ask you to click on a link; this link takes you to a fake 'Sign on' page identical to that of your bank or online shop. The fraudsters use this fake page to capture your details. Here is a fake e-mail I received about my PayPal account:

Dear PayPal Customer,
As part of our security measures, we regularly screen activity in the PayPal system. We recently noticed the following issue on your account.
Your bank has contacted us regarding some attempts of charges from your credit card via the PayPal system. We have reasons to believe that you changed your registration information or that someone else has unauthorised access to your PayPal account. Due to recent activity, including possible unauthorised payments placed on your account, we will require a second confirmation of your identity with us in order to allow us to investigate this matter further. Your account is not suspended, but if in 48 hours after you receive this message your account is not confirmed we reserve the right to suspend your PayPal registration. If you received this notice and you are not the authorised account holder, please be aware that it is in violation of PayPal policy to represent oneself as another PayPal user. Such action may also be in violation of local, national, and/or international law.
PayPal is committed to assisting law enforcement with any enquiries related to attempts to misappropriate personal information with the intent to commit fraud or theft. Information will be provided at the request of law enforcement agencies to ensure that perpetrators are prosecuted to the full extent of the law.
To confirm your identity with us click here:
Case ID Number: PP-731-921-730
After responding to the message, we ask that you allow at least 72 hours for the case to be investigated. E-mailing us before that time will result in delays. We apologise in advance for any inconvenience this may cause you and we would like to thank you for your co-operation as we review this matter.
Sincerely,
PayPal Account Review Department
PayPal E-mail ID PP572

If you receive any e-mail asking for confidential information, do not reply to it. Banks, shops, online business *never, ever* ask for your details in this way. If you are unsure, contact your bank or internet shop directly. I would recommend that you type the web address directly into your web browser or call them.

Remember: criminals are always thinking of new ways to convince you to reveal your details. The latest is to write explaining that you have won a major prize. Because it appears to be from your bank or online shop, you are more likely to believe it. The first e-mail never requests any personal details because they do not want to arouse your suspicion. Once you reply, however, the second e-mail gives you a confirmation of your prize together with a request for you to confirm all your account details.

Now you are forewarned, you won't be fooled!

User names

Many internet shops and businesses require all users to join or become a member. The procedure is nearly the same for all of them in that you create a user ID so they can identify you each time you visit. Try to think of an ID that is easy to remember, that is not too complicated and does not contain your e-mail address or real name. Do not be surprised if your first choice is already taken.

Don't worry about suspicious e-mails

If you receive an e-mail you are unsure of, just delete it. If it is important, the sender will call you or write to you.

Passwords

Your password is linked to your user name and is the only means you have of preventing others from gaining access to your internet accounts. When signing into your account, you will be asked for your user ID and your password. It is really important that you tell no one your password and make sure you change it regularly.

I recommend that your password should:

• Be six to eight characters long;

• Be made up of a mixture of capital letters, numbers and special characters;

• Not include words that can be found in a dictionary;

• Be impossible to guess.

Secret question

In addition, many internet shops and businesses require you to register a reply to a secret question. You will be provided with a list of options for you to choose from: for example, 'What street did you grow up on?' or 'What is your mother's maiden name?' You must select a question and provide your answer, making sure it is a reply that is easy to remember.

Your reminder box

- Never, *ever* reveal your account details – banks and businesses never request them in e-mails.

- Avoid clicking on links within e-mails, as it is much safer to type the address into the web browser yourself.

- Always make your password impossible to guess.

- Any e-mail marked 'Urgent' or 'Respond within 24 hours' should be treated with suspicion. If in any doubt, delete it.

Chapter 4

How to Search for Internet Shops

Within this book, I have listed a variety of places to shop and hopefully I have covered the majority of your needs. However, the internet is a fantastic resource for the unusual, and therefore I would encourage you to search for your own internet shops and services. With hundreds of new businesses appearing online every week, it is really important that you learn how to search for them yourself.

Search engines

You have probably used a search engine before, perhaps not even knowing you have. Search engines simply find websites. You enter a word or phrase and within a few seconds the results are displayed. You then click on the results that closest match your request. The most popular search engine sites include:

- www.google.co.uk
- www.yahoo.co.uk
- www.ask.co.uk

Using a search engine

- Key **www.google.co.uk** into your address bar and click **Go**.

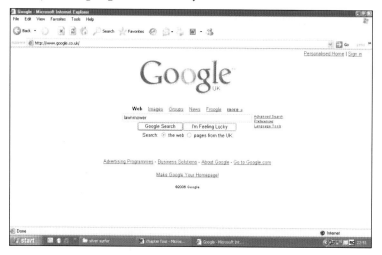

- Now enter **lawnmower** into the search box.

- Click in the circle next to **Pages from the UK** so a dot appears in it.

- Finally, click **Google Search** to begin your search.

Your search results are displayed as a list, beginning with those the search engine thinks are most appropriate to your criteria.

The top results in the shaded blue box and the results on the far right side of the page are sponsored links. Sponsored links are websites that have paid to be featured.

If you look carefully at the results, in the top right-hand corner you should notice that there are over four million results listed: far too many to look through.

In order to find exactly what you are looking for, you need to be really specific about your search criteria. To do this, you need to choose the right keywords to search.

Don't worry **about searching**

Many links provided by search engines are sponsored, so don't worry if what you are looking for does not appear right away. Try using other keywords to find what you are looking for. On Google, the sponsored links appear on the right-hand side.

Choosing the right keywords

To see how much more effective a search can be, try another search.

- Key **www.google.co.uk** into your address bar.

- Enter **petrol lawnmower** into the search box.

This time the results are far fewer, just over 100,000.

Now click on any of the underlined results and a link will take you to the website so you can look at it at your leisure. To go back to the results page, click on the back arrow in the top left corner of the toolbar. You will find the back arrow very useful, as you may have to look at several search results before you find what you are looking for.

Refining your search

Now try another search using keywords or by using an exact phrase. For example, try putting petrol lawnmower into quotes – "petrol lawnmower" – and search again. In this case, Google will only give you sites in which the two words appear together.

For this search, you'll be given just over 16,000 options. It is important you have lots of practice in using keywords and phrases. If you experiment, you will soon get the hang of it. If you narrow your search correctly then you should find what you are looking for on the first search results page!

Don't worry **about 'I'm feeling lucky'**

This search option within **www.google.co.uk** takes you direct to what Google believes to be the most appropriate site based on the keywords you have entered.

Web-friendly sites

There are now websites that are designed with you in mind; they are simple to navigate and personalised, so they remember all the links you have clicked on before. This can save you time – and you don't have to remember where you have been! The site will welcome you each time you visit, and display your favourite internet shops, weather and local news in your area, all on your very own personal welcome page. All this happens without you having to type a thing into your web browser – and it is totally free, too!

• Type **www.virginsurfer.co.uk** into your web browser.

To benefit fully from the site you must register your details. Have a look at some of the links by clicking on them. If you wish to join, click **Register**.

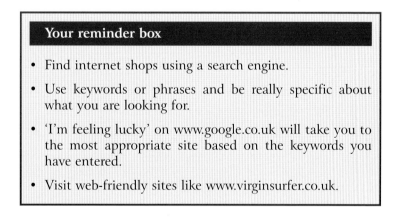

Your reminder box

- Find internet shops using a search engine.

- Use keywords or phrases and be really specific about what you are looking for.

- 'I'm feeling lucky' on www.google.co.uk will take you to the most appropriate site based on the keywords you have entered.

- Visit web-friendly sites like www.virginsurfer.co.uk.

Chapter 5

Price Comparison Websites

Most of us tend to compare prices before purchasing an item. You can, if you wish, search out the best deals yourself on the internet by using one of the many search engines and making a note of the price and where you saw it. It is similar to walking down your high street and making a note of the price in each shop, but you do this from the comfort of your own home.

Another way to compare prices is to visit a price comparison site. These sites check prices across many internet shops and businesses, and may also link to eBay, the internet auction site. We will look at eBay later (see page 131).

Using price comparison websites

Before you start to search on any price comparison site you must follow the advice on pages 32–33 regarding searching and using keywords.

Don't worry about strange company names

Sometimes price comparison sites can provide too many choices and might throw up the names of retailers you have never heard of. If you are unsure, then shop with the trusted high street names you know. You can still save if you compare prices.

When visiting price comparison sites, it works best if you have a good idea of what you are looking for. If you have a manufacturer and a model number in mind, then that's an excellent start.

Comparing prices

In this example, I'll show you how to search for a television. We are going to search for a price for a Philips 42pf5521d television; you will notice I already have a manufacturer and model number.

Many people view products on the high street before visiting the web to find the best price, and it certainly makes sense if you can view and even handle the item you want before you search for the best price using the internet. High street stores will also offer advice and should help you to choose the right product. I've visited them myself to see products I am interested in, and I often revisit them when I've found the best price on the internet to see if the store will match the price.

With electrical items such as televisions and washing machines, it is worth bearing in mind that these items are usually heavy and some require installation. A high street shop can arrange this for you, so you should be prepared to pay a little more than your internet price if delivery and installation are included. I strongly believe in supporting local businesses, so ask your local store if they can offer you a deal once you have a print-out of your best internet price. For now, let's find the best price on our Philips television with the help of a price comparison website.

- Type **www.kelkoo.co.uk** into your web browser.

- Type **Philips 42pf5521d** into the search box and click **Search**.

- The results will be displayed. Click **Go** next to the best price and carefully read all of the description to confirm that it is a new product, to check the delivery charge and establish whether the product requires installation.

Installation is really important, as most products will just arrive in a box and any installation is down to you to sort out. Before you buy anything, it is best to make your own enquiries as to how much this will cost if you are unable to install what you have purchased. Always consider this extra cost when buying on the internet, as your local store could offer you a cheaper deal. There's no harm in asking!

Now try another website.

- Enter **www.pricerunner.co.uk** into your web browser.

- Now type **treadmill** into the search box and click **Search**.

Because you were not specific regarding a manufacturer or model, you now have lots of sites to look through. This is quite helpful if you cannot visit a local fitness store to compare models.

- Click the back arrow in the top left of your screen to return to the PriceRunner homepage.

- Enter **reebok I-run** into the search box and click **Search**.

- You see that now you have located a Reebok I-run treadmill.

Experiment using different comparison websites to look for different items until you are happy with the way they work.

Useful websites

Try these price-comparison websites.

- www.kelkoo.co.uk
- www.shopping.co.uk
- www.pricegrabber.co.uk
- www.pricechecker.co.uk
- www.pricerunner.co.uk

Your reminder box

- If you can, physically look at and handle the items in shops before finding the best price on the internet.

- Use lots of price comparison websites in order to find the best deal.

- If your item requires installation, check the cost first.

- Once you have found the best price on the internet, check whether your local store can match the price.

Chapter 6

Paying for Items from the Internet

It is simple and safe to pay for goods on the internet if you follow the commonsense rules we have established.

There are really only four ways to pay for your purchases from the internet: debit card, credit card, personal cheque and PayPal.

Debit cards

Debit cards are linked direct to your bank account. If you use your debit card, the funds are taken almost instantly from your bank account.

Credit cards

Many people may be surprised to hear that I believe credit cards are the best method for paying for items that you purchase via the internet. I understand that some of you might not own a credit card and that it is essential for anyone to be mindful of their own circumstances before applying for one, as they do hold the potential for you to run into debt. However, they do have advantages over a debit card.

Credit cards provide an excellent level of consumer protection when things go wrong. If your items don't arrive and, having tried to get in touch with the internet retailer a few times, you fail to get a response, then you can simply claim the money back from your credit card company. This is known as a chargeback. The card company will refund the transaction and recover the money from the retailer.

 Don't worry if you don't have a credit card

Most sites now accept debit cards. Stick to the internet sites that have high street stores and you won't have a problem.

Personal cheque

Only a limited number of internet shops will accept payment by personal cheque, and when they do they will always retain the goods until the cheque clears, which can take up to 10 days. It is not the safest or fastest method of payment. Some users of the internet auction site eBay still take this as a method of payment (see page 152).

PayPal

PayPal is a quick and easy electronic method of payment. It was established in 1999 and has been a huge success. It was so successful, in fact, that eBay purchased the business and since then it has gone from strength to strength. As well as eBay, other businesses now use PayPal as a way of accepting payment.

For you to use PayPal as a payment method, you first have to register with them. This process does take a little while but once registered you can make payments directly to other PayPal users.

The system is safe to use as its technology means that all your details are kept secure and confidential at all times. It also offers buyer and seller protection that covers you for up to £500, but you do have to meet some conditions to qualify, unlike a credit card chargeback.

Don't worry **about PayPal**

Although PayPal is extremely useful and safe, it is perfectly possible to shop online successfully without it. But if you want to know more, go to **www.paypal.co.uk** or see page 48.

Using debit and credit cards

When purchasing items on the internet, you will at some point be asked to enter your credit or debit card information. Sometimes this must be entered when you register or become a member. Your card details are then kept on file ready to be used when you make your purchase. This saves you from having to enter your card details every time you wish to make a purchase from the site.

I am always apprehensive about giving my card details to anyone. On the internet you must always check that the internet page you are entering your details into is encrypted and secure. Thankfully, you don't have to be Sherlock Holmes or a computer genius to work this out. Look closely at the right-hand side of your computer screen towards the bottom and you should see a yellow padlock. This tells you that you are on a secure page.

Don't worry **about credit card security**

If you are worried about using credit cards over the internet, simply apply for another card and ask the card provider to set a low credit limit for you. You can also do the same with debit cards by opening a bank account just for internet use in which you deposit only a small amount of money to cover your transactions.

The card details commonly requested are:

- Card type;

- Card number;

- Start and expiry date;

- Security code/CCV code – this is the last three digits on the back of your card;

- Issue number, for some debit cards.

---

b) Payment card details

Name on card: *

Card number: *

Expiry date: *   -- / --

Issue Number:

Security code: *

This is a unique number printed on your card. On most UK cards this is the last three digits of the number printed on the signature strip on the back of the card. Find out more about security codes.

(Note - we never store your Card Security Code)

🔒 🌐 Internet

Your reminder box

- A credit card is often the best way to pay for internet transactions.

- PayPal is a quick and convenient payment method, particularly for eBay purchases.

- Look for the padlock symbol before divulging your credit card details.

- Payment may also be made by debit card and personal cheque.

Chapter 7

Using PayPal

If you are planning to buy goods through eBay, it makes sense to open a PayPal account. This is very easy to do and it is free to register.

Registering with PayPal

- Type **www.paypal.co.uk** into your web browser.

- Click **Sign Up Now**.

- You will be offered the choice of a number of different accounts, but we are only really interested in one of two accounts. Read the following information, then select the account you require.

- **PayPal personal account:** This is the best option for beginners. If you are only interested in buying – rather than selling items on eBay – select this account. It is perfect for anyone wanting to buy items from the internet, including eBay. You can upgrade to a premier account at any time if you change your mind.

- **PayPal premier account:** If you want to buy and sell, this account gives you the flexibility to be able to receive debit and credit card payments for items you offer for sale on eBay, as well as being able to pay for the items you have bought. Many of the current eBay members have a premier PayPal account. At this stage you might not know much about eBay and have no desire to sell things yourself. However, I know many people who once thought this way and are now real eBay fans, buying and selling regularly. We will look at eBay later (see page 131).

- Select the county or region where you live, then click **Continue.**

- Now, you must complete the account form by filling in all the boxes. Follow my advice on passwords on page 29. Please make sure your telephone number is correct: PayPal will check this detail later.

- There are some additions to the user agreement that you must agree to by ticking **Yes**.

- To prevent automated registrations, the additional security measure of a random text box has been introduced. Simply type the characters that appear into the blank box on the right, then click **Sign up**.

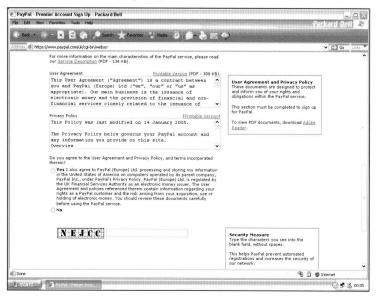

Don't worry **about PayPal**

Don't worry if you find the PayPal system a little complicated at first glance. You will understand more when we have a look at the internet auction site eBay later in this book. For now, you don't need a PayPal account to shop on the internet.

Confirming your PayPal registration

- PayPal will send you an e-mail with a link asking you to activate your account. Click on the link to take you to the PayPal site.

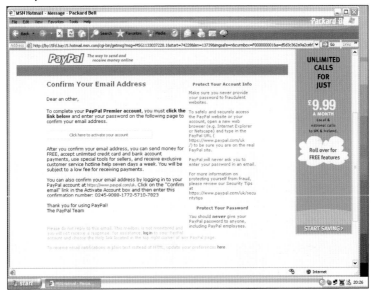

- Enter your password, and your PayPal account is up and running.

Your PayPal account

Once you have registered with PayPal, you can choose to pay for goods through your PayPal account, providing the internet shop or eBay seller accepts this method.

If you become an eBay seller, you will soon build up a balance of cash within this account. Use this balance for paying your eBay fees, and transfer amounts of £50 or more to your bank account to avoid transaction fees. Don't keep lots of money in your PayPal account as no interest is paid on the account.

PayPal fees

PayPal offers a great service that is completely free for buyers, so if you only use your PayPal account to buy goods, you will not pay anything.

PayPal does charge a small commission if you are a seller. These fees will alter from time to time so if you are selling items, you should always check the most up-to-date charges for yourself by visiting www.paypal.co.uk.

Using your PayPal account

Each time you visit www.paypal.co.uk and log in to your account overview, this page is displayed, showing your recent activity.

On the top of the page you will see a number of tabs linking you to the services available. We will explore the more useful of these services throughout this chapter.

Funding your transactions

Each month, many millions of pounds' worth of transactions are handled by PayPal. To prevent fraud, PayPal has introduced a number of checks to confirm your identity and your address.

All new accounts have a spending limit of £500. If you are an eBay seller, there is no limit to the money you can receive into your PayPal account. However, there is a withdrawal limit of £500 per month for a period of three months.

If you wish to send money or purchase items and pay for them via PayPal, you must now register a credit or debit card to fund your transactions. This can be added during the verification process.

The verification process

You will be asked from time to time to verify your account. Once verified, your spending limit will be lifted. The only limit you then have is the balance on your chosen method of payment: your debit or credit card.

There are three steps you must complete for your account to be verified:

- Go to **www.paypal.co.uk** and sign in to your account.
- Select the **Get verified** link.

Stage 1: adding your bank account

- Select **Set up bank funding**.

- Choose your country from the drop-down list, then click **Continue**.

- You must now fill in all the boxes and reconfirm your banking details. Once you have completed this, select **Add bank account**.

- PayPal will credit two small amounts into your bank account. This usually takes between three and five days. When you see your next bank statement, make a note of these amounts – they will be needed later.

Confirm your bank account and set up a direct debit

- Now access your PayPal account by going to **www.paypal.co.uk** and signing in.

- Fill in the two boxes with the two small amounts that were credited to your account. This will verify your bank account and you will have completed the first of PayPal's three security checks.

Add your credit or debit card and validate your account information

- The next step is to add your credit or debit card details. Once logged in to your PayPal account, select the **Get verified** link from the left-hand menu.

- If you have been successful in setting up and validating your direct debit, a tick will appear in the first box, confirming that the first of the three steps has been completed.

- Now select **Validate your account information**. Take your time in filling in all the boxes with your credit or debit card information.

- Once you have completed the boxes, select **Add card** at the bottom of the page and the account overview page will appear.

Don't worry **about the verification process**

PayPal's three-stage verification process may seem a nuisance, but it does offer you a high level of security on your internet transactions, so it is well worth it.

Stage 2: validating your account by phone

In the next step, PayPal will ask you to check that the address and home telephone number you registered with your debit or credit card are correct. To help combat fraud and protect its members, PayPal makes an automated phone call to the telephone number entered for your address, so please ensure the details you have entered are accurate. You need to have a touch-tone telephone (one on which you hear tones when the buttons are pressed).

• Go to **www.paypal.co.uk** and sign in to your account.

• Select the **Get verified** link.

- At the bottom of the website page, you can choose to receive the automated call immediately or after one minute by selecting the appropriate link. Before you click **Continue**, make sure that no one is using the telephone and that you have the phone within reach.

- Once you have clicked **Continue**, PayPal will issue you with a PIN.

- You will shortly receive an automated call from PayPal. When you answer this call, you will be asked for the PIN displayed.

- After you have entered this PIN using the telephone keypad, you can simply hang up.

Congratulations! You have successfully validated your account information and completed stage two of the verification process.

Stage 3: edit business information

This final stage is very simple and is only there because PayPal has a legal requirement to know a little about you.

- Go to **www.paypal.co.uk** and sign in to your account.

- Select **Get verified**.

If you have been successful in validating your account details by phone, a second tick will appear in box two, confirming that this second stage has been completed.

- Now select **Edit Business Information**.

- You will be presented with the final form to complete. It is simple, so try to answer all the questions as best you can.

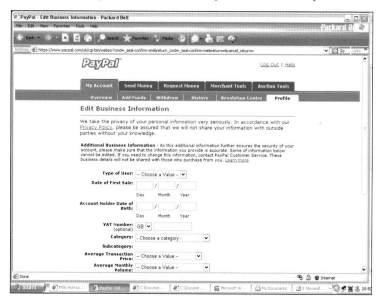

- When you have finished, select **Save** and you will have completed the verification process.

Adding funds

There are really only two ways in which money is transferred into your PayPal account. One is if you sell an item – on eBay, for example – in which case a buyer will send you money via the PayPal system. The second is that you transfer money yourself from your registered bank account.

Funds from a buyer

When a buyer sends you money via the PayPal system, you will receive an automated e-mail stating, 'You've got new funds!' Within the e-mail will be a note of the amount and who has sent it. There is also a link that will show you the details.

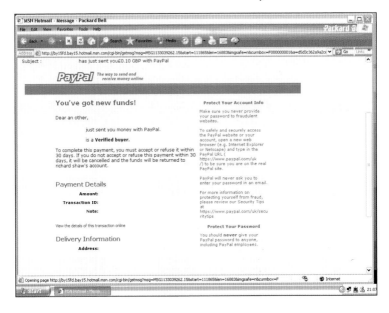

Funds from your bank account

You may at times want to transfer funds from your bank account into your PayPal account. It is a simple process and is free, although do remember that no interest is paid on your PayPal balance and the process usually takes seven to nine working days. It is a good idea to check the amount before requesting a transfer into your PayPal account, and always make sure you have the funds within your bank account to cover the amount.

- First log in to your PayPal account, then select **Add Funds**.

- Now select **Transfer funds from a bank account** and follow the straightforward instructions.

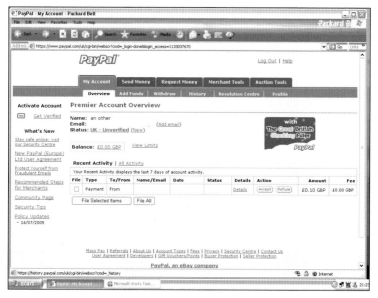

Paying through PayPal

If you are buying goods on the internet and you go to the checkout, the company may offer you the option of paying by PayPal. If so, click on the link and it will take you through a step-by-step process to pay for your goods.

If you have won an item on eBay, you will be sent a 'Congratulations' e-mail giving details of the item you have won and payment options accepted by the seller.

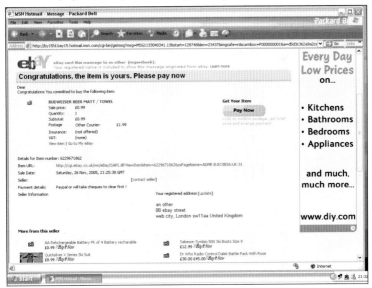

- Select **Pay Now** within the e-mail as your payment method, and check all the details are correct. The link will take you to the PayPal site.

- You will be asked to log in to your account to confirm who you are, then click **Continue**.

- Next, review all the payment details to ensure that you are happy with your funding options, then click **Continue**.

That's it! You have made your first payment! An e-mail will then be sent to you confirming that a payment has been made.

Don't worry **about being paid by PayPal**

This is only relevant if you become a seller with a premier account, rather than a buyer.

Sending money

There might be occasions when you would like to send money for non-auction goods and this can be done via PayPal. You can even use PayPal to send money to a friend or relative for their birthday, for example (if they also have a PayPal account) – particularly useful for overseas.

- Go to **www.paypal.co.uk** and sign in to your account.

- Select the **Send money** option.

- Fill out the boxes, including the recipient's e-mail address, then select **Continue**.

- Make sure the payment details are correct. If you wish to change your funding options, select the **More funding options** link found in the **Source of funds** section.

- When you are happy that everything is correct, click **Send money** and you are finished.

Resolving payment disputes

Mistakes do happen, so if you ever have a problem with a purchase, contact the seller first. Remember: co-operation is better than confrontation. Only when you have done this and you find yourself making no progress should you use PayPal's resolution centre.

- Go to **www.paypal.co.uk** and sign in to your account.

- At the top of the page, click **Resolution Centre**.

Then you simply follow the straightforward procedure to log your claim. You will receive an e-mail acknowledging your dispute information, followed by e-mails to let you know what has been done and how the dispute has been resolved.

Remember: if you have paid for something using a credit card you may contact the card company and make a claim.

Your reminder box

- The most common method of payment on the internet is by credit or debit card.

- If you used a credit card for payment and things go wrong you can make a claim with your card company. This is called a chargeback.

- PayPal is used mainly on the internet auction site eBay and by some businesses.

- You do not have to have a PayPal account to shop on the internet.

Chapter 8

Shopping for Groceries

There is nothing new in having groceries delivered. I can remember my older brother making deliveries on a pushbike with a large wicker basket on the front. What has really changed is the price and choice.

From the comfort of your home, you have a choice of thousands of different items from your favourite stores and you can have them delivered directly to your door. Ask around to see if anyone you know has experience of shopping for groceries online, because price is not the only consideration. Read the site's small print to determine:

- Are the goods picked from a supermarket branch or a warehouse?

- How precisely can you choose your delivery slot?

- What happens if you miss the delivery?

- What is the policy on substitution? If you order something that is not available, what will the supermarket send in its place? Do you have any say in this? What is the company's record on substitution?

If you live in the catchment area of one of the big supermarkets, you can enjoy timed deliveries and lots of choice. Nearly all internet grocery websites will ask for your postcode to determine from the start if they can deliver to you.

If you are on a budget, internet supermarkets can be a great way of assisting you in purchasing just what you need. The act of placing an order via the internet encourages you to think really hard about your meals and groceries, avoiding all the temptations in store that lead to you spending more money than you planned.

Registering on a supermarket website

Now visit a supermarket that has a big internet presence.

*   Type **www.tesco.com** into your web browser.

As you can see from Tesco's home page, it offers more than just groceries. As with most internet supermarkets, it offers a huge range of products and services, including finance, insurance, books, electrical equipment and even fuel supplies.

*   In this exercise, you are interested in having your groceries delivered to your door, so click **Groceries** from the options near the top of the screen.

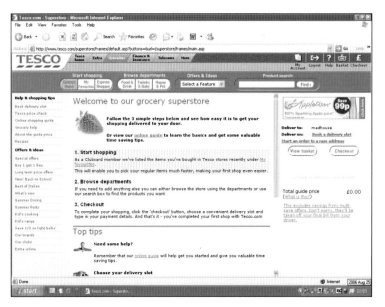

Within this page you have two options:

- If you are an existing customer you must sign in with your account details.

Don't worry **if the site changes**

Or we probably should say *when* the site changes. The layouts on major sites will be updated on a frequent basis, so don't expect them to remain identical each time you visit. However, the basic instructions and sequences will remain the same.

- If you are a new internet customer then you must register your details. Complete the **Register here** box.

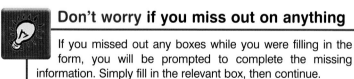

- Once you have entered all your details, click **Register**.

- You now have another short form to complete. Towards the end of the form, you will be prompted to read the Terms and Conditions and confirm you understand them by clicking on the **Yes** button.

Don't worry **if you miss out on anything**

If you missed out any boxes while you were filling in the form, you will be prompted to complete the missing information. Simply fill in the relevant box, then continue.

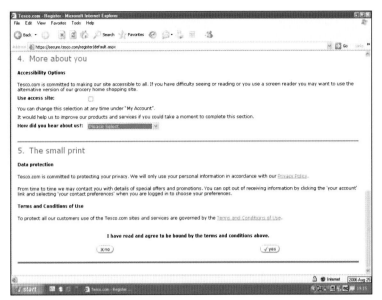

- If all the boxes have been completed then you will be taken to a 'Thank you' page. Click **Continue**.

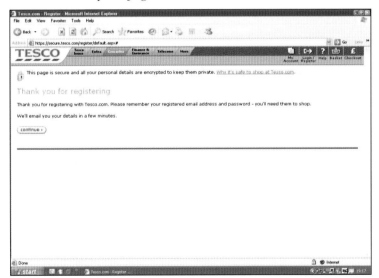

A welcome e-mail will also be sent to the e-mail address you have provided.

What to expect from a supermarket website

You are now signed in to the Tesco website. There are many aspects of this site that are common to nearly all shopping websites. Let's take a closer look at the icons in the top right-hand corner.

My account

This takes you to your account details.

Basket

This takes you to a page displaying all the products that you have so far selected for purchase.

Checkout

This takes you to the payment page. Here you can check that all the details are correct. Only click **Confirm** when you are completely happy with the items you have chosen, the delivery time and date and the total cost.

Log out/sign out

Whenever you have finished shopping, it is advisable that you click **Log out** or **Sign out**. This ensures that no one else can continue shopping using your computer and account details.

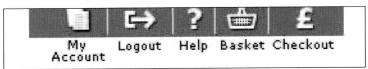

What you will find on the groceries page

To start shopping, select **Groceries** at the top of the screen.

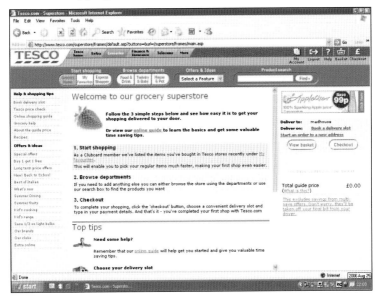

If your address falls outside Tesco's catchment area or is registered to a business, then you will be unable to shop for groceries via the internet. However, you may fall into the catchment area of another supermarket, so it is worth checking the websites of Adsa or Waitrose, for example.

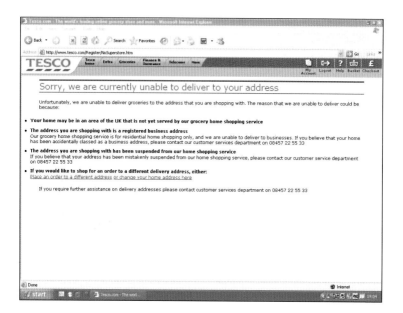

Have a good look at the screen. Again, there are lots of features common to most internet shops. Don't worry if it appears complicated; just take it one section at a time. Start with the Tesco toolbar.

Grocery home

The **Home** button or tab will always take you to the grocery home page.

My favourites

This takes you to your favourites page, where you will find links to:

- All my favourite items
- My last order
- Toiletry and baby favourites
- House and pet favourites
- Delete favourite items.

Express shopper

When you go to the Express Shopper page, you can type your shopping list within the notepad on the left of screen. As with the search engines you have used before, you must be as specific as you can. Once you have written your list, click **Find now** and the site will find all your shopping for you.

Food and drink

This takes you to the Food and Drink page from where you can choose items to purchase.

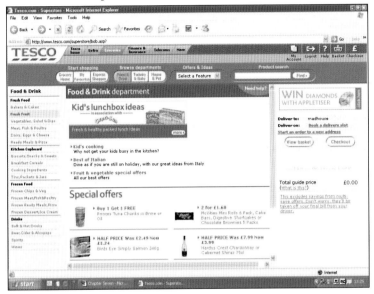

Offers and ideas

Most supermarkets will remember your weekly order and, to make life simpler, allow you simply to repeat it. You may find this useful if you have the same meals every week. However, always check out this page for special offers, such as 'buy one get one free' before placing your order. Displayed on the far left are a number of categories, many of which are divided into subcategories to make items easier to find.

Food & Drink
Fresh Food
Bakery & Cakes
Fresh Fruit
Vegetables, Salad & Dips
Meat, Fish & Poultry
Dairy, Eggs & Cheese
Ready Meals & Pizza
Kitchen Cupboard
Biscuits, Snacks & Sweets
Breakfast Cereals
Cooking Ingredients
Tins, Packets & Jars
Frozen Food
Frozen Chips & Veg
Frozen Meat, Fish & Poultry
Frozen Ready Meals, Pizza
Frozen Dessert, Ice Cream
Drinks
Soft & Hot Drinks
Beer, Cider & Alcopops
Spirits
Wines

Select the drop-down menu to search for 'buy one, get one free' and other special offers.

Product search

This is like a search engine within the Tesco site. Use this search box to look for any item, but remember to try to be specific about what you need.

How to shop for groceries

This will give you simple step-by-step instructions on how to do your shopping at Tesco. All the other supermarket sites will work in a similar way.

- Go to **www.tesco.com**.

- Select **Grocery**.

- Once on the grocery page, click **Food and drink**.

- Now select **Fresh Fruit** from the list on the left.

- Then click **Bananas** from the fruit list.

- You can see from the list that there is more than one type of banana. For this example, we are interested in the cheapest loose ones. Click the + arrows until you have 10 bananas and select **Add**. You also have an opportunity to add a note. For example: 'I don't want any green ones, thank you.'

By clicking **Add**, you are placing these bananas in an imaginary shopping basket. To view this basket and remove items, click **View basket**. Next to this you will see your basket total.

Deliver to:	madhouse
Deliver on:	**Book a delivery slot**

Start an order to a new address

(View basket) (Checkout)

⊖ 10 ⊕ Bananas Loose 1.50

Total guide price **£1.50**
(What is this?)

This excludes savings from multi-save offers. Don't worry, they'll be taken off your final bill from your driver.

When you are actually doing your shopping, you simply carry on searching items and adding them to your shopping basket until you have done your weekly shop. Most supermarket websites will set a minimum spend level. For the purposes of this exercise, assume that you are on a banana diet and have finished your shopping.

- To proceed to the checkout, simply click **Checkout** to bring up the delivery page.

You will now be asked when you would like your groceries delivered. With Tesco, you are able to book a two-hour delivery slot, which saves you waiting in all day for the delivery. Tesco makes a small charge to deliver your groceries. You can take advantage of the cheaper time slots to save yourself money. It may not be much individually, but if you shop online regularly, the savings can soon add up.

- Click your chosen time slot to continue.

Paying for your shopping

- You will now be asked to confirm your order and enter your payment details.

- For the purpose of this exercise, we will leave without confirming the order. Click **Logout** at the top of the screen to exit.

Don't worry **about confirming your order**

An order is only placed when you have confirmed your payment details and then clicked **Confirm my order.**

If you click **Confirm my order**, your order is placed and you are taken to the receipt page where, if you wish, you can print off a receipt for your shopping.

More grocery shopping websites

By using internet search engines you should be able to find all the internet grocery shops and supermarkets you could ever wish for. Here are some of the most popular:

- **www.asda.co.uk** Asda, like Tesco, sells just about everything. It aims to offer the lowest prices.

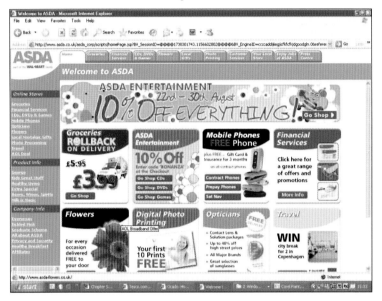

- **www.waitrose.com/www.ocado.com** Waitrose tends to offer the more unusual items from smaller suppliers, as well as offering a good selection everyday items.

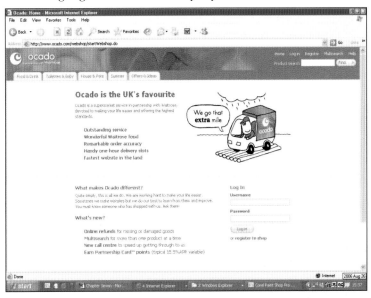

 ## Don't worry **about surfing**

You can spend as long as you like finding your way around websites you visit. There's no pressure to buy, and you can go back as often as you like.

- **www.sainsburys.co.uk** Sainsbury's has a good selection of everyday items at competitive prices.

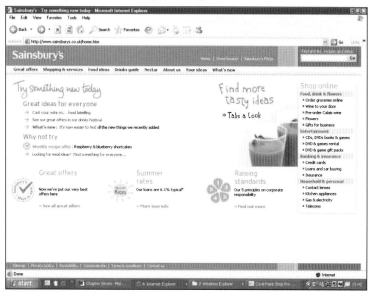

Your reminder box

Most internet shopping sites share the same features as the Tesco website described on pages 66–81. These are:

- Register your details section;

- Sign in/out;

- A search box;

- A shopping basket;

- A checkout;

- A 'Confirm/Complete' link to place your order.

Don't forget that price is not the only factor to consider.

Chapter 9

Household Goods and Services

Just as with grocery shopping, buying household goods and services online can save you time and money – not to mention the hassle of carrying stuff home. But take your time, do your research and don't forget to include the cost of delivery in your calculations before you click to confirm your order. Note, too, that some sites automatically add a nominal sum for insurance of your goods while in transit.

Another point worth noting is that once you have purchased something via a website you will probably be sent e-mails telling you about special offers and other information that the company thinks might be of interest to you. It is easy to stop these: at the end of the email there will be a hyperlink to an 'unsubscribe' service.

Shopping for electrical goods

Before you begin to shop there are a few things you need to consider.

- Do you know what you are looking for? Make a list of the features you would like so you can make sure you are making sensible comparisons. Use the information on websites – printing it out if it helps – to compare both prices and specifications.

- Are the delivery charges reasonable?

- Will you be able to move the item when it arrives? Most items will arrive by courier and they will not help you unpack it or move the product very far.

- Can you install the item yourself or will you have to arrange locally for someone to come and install it? If so, how much is this going to cost?

- Will it fit into the space you have in mind for it?

As mentioned before, it is always best, if possible, to view and handle products before you buy them from an internet shop. If you are comparing prices, you will want to check out your local retailers anyway, so that's the perfect opportunity.

Don't worry about taking your time

You can visit as many sites as you like as often as you like before you make your final decision. Print out information, such as specifications, if it helps.

It would be impossible for me to give specific advice on every electrical item you will ever need to buy, but let's look at the following major purchases.

Washing machines

If you are simply replacing an existing washing machine, then you might be able to fit it yourself, as most modern machines simply use screw fittings. If you are unable to install it yourself then you must include the cost of the installation when you make your price comparison between a high street store that will offer you installation, and an internet shop that probably won't.

Key features

As all washing machines look very similar, you do need to look at the list of features when shopping on the internet so you can make sure you are getting what you want. Here are the features you should be looking for:

- **Washer/dryer:** Is it a combined washer/dryer or just a washing machine?

- **Load capacity:** If you do a lot of washing, then a machine with a large capacity would be a good choice.

- **Spin speed:** The faster the spin, the less the drying time.

- **Quick wash:** A fast wash programme if you are in a hurry.

- **Hand wash:** A wash programme for 'hand wash' items.

- **Half load:** Allows you to wash smaller loads and therefore save money.

- **Ratings:** All washing machines should have official energy-use ratings. They are graded from A to E, with A being the best. If possible, you should look for energy A, wash rating A and spin A. This is known as triple A (AAA).

Dishwashers

Just like washing machines, if you are simply replacing an existing dishwasher then you might be able to fit it yourself, as most modern machines use screw fittings. If you are unable to install it yourself, then you must include the cost of the installation when you make your price comparisons.

Key features

- **Place settings:** How many place settings can the dishwasher clean at once? If there are only two of you, a small dishwasher may be better, but if you do a lot of entertaining you might want a bigger one.

- **Wash programmes:** How many wash programmes does it have and what do they do? Before you go for anything complicated, think about the fact that most people only ever use the ordinary wash cycle.

- **Economy wash:** A money-saving wash programme that uses less water and energy.

- **Half load:** Allows you to wash a smaller load and therefore save money.

- **Ratings:** Like washing machines, you should be looking for an AAA rating, this time related to cleaning, energy and drying performance.

Fridge freezers

The only real advice I can give you is to make sure it fits the space you have in mind for it, and make sure it is transported upright. If you have any doubt that it has, then you must wait the manufacturer's recommended time before you plug it in.

Key features

- **Capacity:** Based on your own needs, decide whether the freezer or the fridge part should be the larger and how much capacity you require.

- **Frost free:** You never have to defrost again or face mopping up a wet kitchen floor.

- **Ratings:** Energy rating is important; you should be looking for A.

Televisions

In the last couple of years, LCD and plasma televisions have really taken off. These televisions are the slim, flat screen ones and have a few advantages over the normal bulky televisions we have been used to watching to date. LCD televisions have a low power consumption and are therefore cheaper to run. Because they are light and slim they can save you space, too, and many people choose to mount them on the wall.

Key features

Technology is changing all the time, and if you are going to spend a lot of money on a television you want to be sure it is going to be as future-proof as possible. Consider a television that has the following features:

- **Freeview/digital:** This allows you to receive over 30 channels if they are available in your area.

- **HD ready:** High definition (HDTV) is the next generation broadcast format. It offers two to four times more picture detail compared to today's picture quality. To watch HDTV you need a TV that is HD ready and an HD box. This HD box connects to your HD-ready television via the sockets on the back. In the future I am sure that the BBC will offer HD broadcasts, but for now your HD box is available on subscription from a satellite or cable supplier.

- **Scart sockets:** These are the sockets on the back of the television that allow easy connection of other equipment such as games consoles, VCRs, DVD recorders and players.

Computers

It is really important before you start searching for a computer that you decide exactly what you want. You will need to spend some time looking at the options available and narrowing them down. This is an expensive item, so do your research.

The first decision is whether you want a desktop or a laptop. Desktop computers tend to be cheaper, but notebooks/laptops are lightweight and so offer portability. They can be used in any room and can even be taken everywhere you go.

The latest computers have a short shelf life before being replaced by a newer, faster model. Unless you have specific requirements for the most powerful computer, it is a good idea to look at the latest offers. These are likely to be on the model that has only just been superseded so will be perfectly adequate.

Don't worry **about laptop keyboards**

Some people find laptop keyboards more difficult to use than conventional ones. If you do find your laptop keyboard awkward, go to a computer store and have a go on an ordinary keyboard. If you find it easier, you can use one with your laptop, connecting via the USB socket.

Keep your budget in mind. It is easy to be persuaded to spend more than you need to.

Key features

Most packages contain the computer, the monitor, the keyboard and the mouse. The most important specifications tell you how fast the computer works and how much memory it has. The specifications are constantly changing so look at the basic packages first and compare them with the more expensive options.

- **Processor:** This controls the speed at which the computer works. It is measured in megahertz (MHz) or gigahertz (GHz).

- **Hard disk or drive:** This tells you how much information the computer can store and is measured in gigabytes (GB).

- **Random Access Memory (RAM):** This affects how many things the computer can do at once and is measured in megabytes (MB).

- **Monitor:** Most will offer a 17 in flat-screen monitor.

- **Operating system:** This is the program that actually runs the computer and will be the latest Windows system.

- **DVD rewriter:** You need this to copy data to CDs or DVDs.

- **USB connections:** These are the sockets into which you plug equipment such as printers or cameras. The more, the better – most new computers have six.

Also think about any extras you may need, such as a printer or scanner. Don't buy a package that contains lots of peripherals if you already have them and they work perfectly well.

Buying a computer

- Type **www.sony.co.uk**.

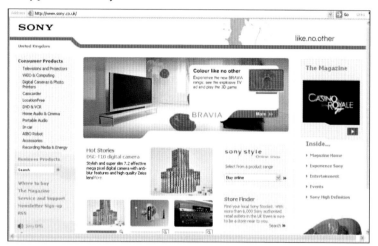

- Now select **VAIO & computing** at the top. In this example we are looking for a notebook/laptop, so select the appropriate link.

- If you wanted to purchase this, then you would select **Buy online**.

- Then add to the basket and proceed to the checkout in the normal way.

Gardening

You can make your garden stand out from the crowd with many thousands of products available from the internet, from plants and seeds to equipment and garden furniture. Due to import regulations, you cannot purchase plants or seeds from some countries. It is also worth bearing in mind that seeds transport better than plants.

Buying seeds

On the left side you see the range of products available. This is a typical website with similar features to those in the Tesco example. In your own time have a good look around the web.

Books

The internet has become a fantastic resource for purchasing rare books together with all the current literature you could wish for. Sites usually have a built-in search engine, so if you only know the title, but not the author (or vice versa) you should still be able to track down the book you want.

• Type **www.foulsham.com** into your web browser.

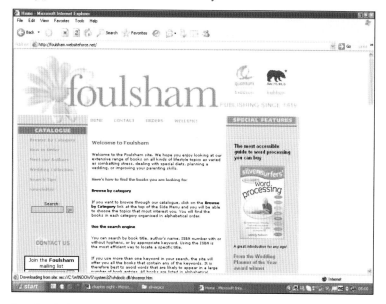

In the left-hand panel you will see the option to browse by category or to use the search box. There's also advice on the most efficient way to use the search engine.

The most popular online bookshop is **www.amazon.co.uk** (or **www.amazon.com** for those outside the UK).

Gas and electricity

Not so many years ago you only had one choice for your gas and electricity – or perhaps two, if you count take it or leave it! Things have changed dramatically, with many companies competing for your custom now. For the best deals, it helps if you have all your electricity and gas needs from the same company and pay for them by direct debit.

If you have never switched fuel companies, it might be worthwhile having a look at the options and most suppliers make it easy for you to switch via the internet. It is very important that you know how much you are currently paying, so before you start searching for discounts have your utility bills to hand. Use a price comparison website (see page 37) to make this easier.

- Type **www.uswitch.com** into your web browser.

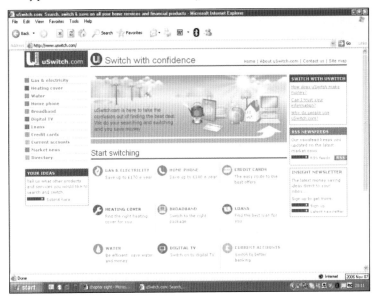

- You will see that Uswitch offers a wide variety of services. For now, click **Gas and electricity**.

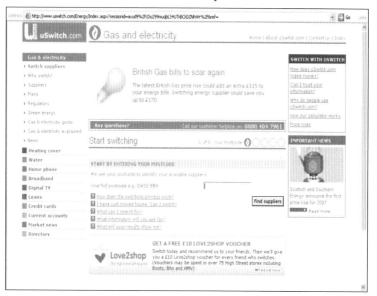

There are only five steps to changing your supplier:

- Enter your postcode to determine available suppliers;

- Enter your current supplier details including how much your current bills are;

- Enter how you would like to pay;

- Choose the best deal for you;

- Enter your bank details.

It really is that simple. However, do seek advice before you switch. Contact your current supplier and ask if it can match or better the deal you have found. Also consider fixed price plans, especially if you are on a budget. Age Concern provides information about coping with fuel bills.

Insurance services

If you just type **insurance** into your search engine you will be faced with a huge list of providers, so you need to refine your search to the specific type of insurance you are seeking. The main insurance services are home and contents, car, life, travel, pet and health.

First visit the well-known companies that regularly advertise on radio and television. Many will give you a summary of the cover they provide, together with the excess payable upon a claim. Some supermarkets also offer insurance now, but compare the terms carefully as just because their fruit and veg are a bargain doesn't mean their insurance will be. Price comparison websites (see page 37) can be useful when you are shopping around for insurance.

Many firms offer a discount if you buy your insurance online, and there may be further reductions if you are over 50. There are also a number of insurance providers that can offer special cover, including travel insurance, even if you have medical conditions. Again, Age Concern offers useful advice.

Useful websites

Electrical goods

- **www.apollo2000.co.uk** Electrical superstores.

- **www.comet.co.uk** Electrical superstores.

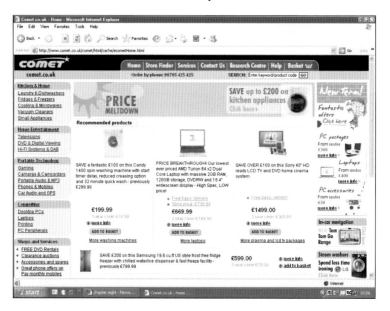

- **www.dixons.co.uk** Electrical superstores.

- **www.electricaldiscountuk.co.uk** Online electrical store.

- **www.hispek.com** Online electrical store.

- **www.hughesdirect.co.uk** Online electrical store.

- **www.simplyradios.com** Online radio sales and repairs.

- Don't forget the supermarkets and **www.argos.co.uk**.

Computers

All these sites offer computer advice, information and sales.

- www.apple.com
- www.computershopper.co.uk
- www.dabs.com
- www.dell.co.uk
- www.ebuyer.co.uk
- www.hp.com/uk
- **www.pcadvisor.co.uk** A useful site providing good computer information in plain English.

- www.pcworld.co.uk
- www.simply.co.uk
- www.sony.co.uk

Gardening

- **www.chilternseeds.co.uk** Seed suppliers.

- **www.diy.com** B&Q's website.

- **www.lawnmowersdirect.co.uk** Online gardening equipment.

- **www.rareplants.co.uk** Online sales of bulbs and rare plants.

- **www.shedstore.co.uk** Sheds and garden buildings.

- **www.simplygardeningtools.co.uk** Online gardening tools.

- **www.suttons-seeds.co.uk** Seed and plant supplies.

- **www.thompson-morgan.co.uk** Seeds, plants and gardening equipment.

- **www.watergardening-direct.co.uk** Products, plants and supplies for ponds and water gardens.

Books

All these sites offer online book sales.

- **www.amazon.co.uk** and **www.amazon.com**
- **www.audiobooks.co.uk**
- **www.barnesandnoble.com** US-based store.
- **www.bookshop.blackwell.com**
- **www.foulsham.com**
- **www.waterstones.co.uk**

Bottled gas and coal

- **www.calorgas.co.uk** Gas supplier's site.

- **www.cpldistribution.co.uk** Solid fuel merchant and home delivery supplier.
- **www.solidfuel.co.uk** Solid Fuel Association website.

Gas and electricity

- **www.ageconcern.org.uk** Charity site; includes advice on energy supplies.

- **www.energyhelpline.com** Information on energy supplies and prices.

- **www.energywatch.org.uk** Gas and electricity watchdog.

- **www.house.co.uk** British Gas website.

- **www.npower.co.uk** Energy company website.

- **www.powergen.co.uk** Energy company website.

- **www.uswitch.co.uk** Offers free online and phone-based comparison and switching service.

Insurance

All these sites offer information and insurance options.

- www.50plusdiscount.com

- www.age-matters.org/ukcarinsurance

- www.rias.co.uk

- www.saga.co.uk

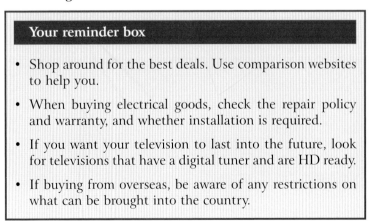

Your reminder box

- Shop around for the best deals. Use comparison websites to help you.

- When buying electrical goods, check the repair policy and warranty, and whether installation is required.

- If you want your television to last into the future, look for televisions that have a digital tuner and are HD ready.

- If buying from overseas, be aware of any restrictions on what can be brought into the country.

Chapter 10

Shopping for Clothes

The internet can bring you all the clothes you will ever need, from everyday wear to top designer labels. What's more, you can save money compared with shopping on your high street. Many of the shops require you first to register your details and set up an account, in a similar way to that required for online grocery shopping (see page 66).

Types of account

Some internet shops will set you up with a credit account allowing you to shop and pay a minimum amount each month. If you can, it is always much better to pay for your items in full. The interest rates charged are much higher than those of the banks and you could find yourself trying to pay off a large bill that attracts a high interest rate. When paying in full, make sure you allow time for your payment to reach the shop and be processed, otherwise you could be paying interest too.

Size guides

Most internet shops you deal with will have a size guide on their site and some might even post you one.

Size guides are really important as you cannot try the clothes on like you can in a store. Note that continental shoe sizes are different from British ones, and American and British dress sizes are different, too, so make sure you know which system is being used before you order.

Returns

Just like in the high street, the actual size of the clothes varies from shop to shop. But unlike high street shops you can't try on the clothes until you have ordered them and they have arrived. With this in mind, it is really important that the internet shop you deal with has a free returns policy.

Most internet shops do offer free returns. They understand that some items simply may not fit correctly or you just don't like the way the item looks on. Items are usually returned by you calling the shop to arrange a collection or by taking them to your local post office to return them with a postage paid label. Most companies will give you up to 14 days to return unwanted items. However, this varies, so always check the return policy before ordering.

Ordering clothes online

Follow this example and you'll soon see how clothes retailers'
sites work. They are all very smilar.

- Type **www.next.co.uk** into your web browser.

- Now click **Womenswear** from the left-hand menu, followed
by **Shoes**, then **Boots**. In this example we going to shop for
a pair of size 6 black boots.

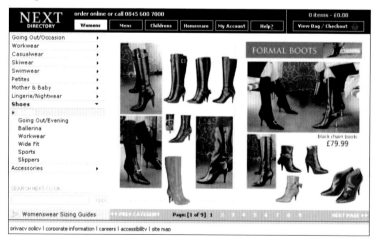

- You can now explore a variety of boots. When you are happy with your selection, click **Add to bag**, remembering to specify that you want size 6.

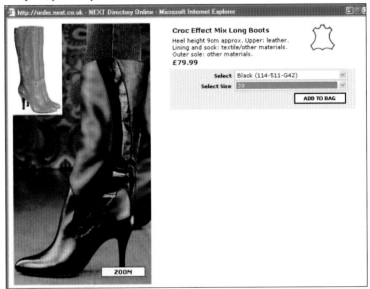

Remember: if you are ever unsure of the size, always check the website's size guide, usually accessed from the home page.

| FAQ's |
| Contact Us |
| How To Shop Online |
| Accessibility |
| Account Information |
| Delivery Information |
| Head Office Information |
| Security & Privacy |
| Site Map |
| Next Electric |
| Sizing Guides |
| Womenswear |
| Menswear |
| Childrenswear |
| Homeware |
| Store Search |
| Technical Help |
| Terms & Conditions |
| Close Window |

SHOE SIZE CONVERSION CHART

continental	35½	36	37	37½	38	38½	39	40	41	42	43
U.K.	3	3½	4	4½	5	5½	6	6½	7	8	9
slipper sizes	S	S	S		M	M	M	M	L	L	

HEEL HEIGHTS

Heel heights are taken from centre back of heel (see diagram) and are only intended as an approximate guideline to assist you in choosing a suitable height of shoe.

FOOTWEAR COMPOSITION GUIDE

- Now click **Go to checkout**.

- You will be asked to sign in. If you already have a Next account, simply type in your details. Otherwise, to open an account click **Register now**. From there, simply follow the onscreen instructions on how to pay and arrange delivery.

Now explore another clothes website.

- Type **www.debenhams.com** into your web browser.

Explore the departments by clicking on the tabs at the top. You will notice it very similar to the previous example. As with most sites, you have a shopping basket and a checkout.

Buying clothes on eBay

If you purchase clothes from an internet store, in general if you are unhappy with your purchase then you may return it, providing you haven't worn it and it is within the return policy of the store. With eBay, you bid on items based on the item description and maybe a photo.

Don't worry about eBay

Buying on eBay is easy, once you get the hang of it. There is more about this subject later (see pages 131-187).

If you place a bid on a size 14 dress and you are the winner, then you pay the seller and the dress is posted to you. If you find when it arrives that the dress is too small or too big or is not in the condition that you were expecting, there is little you can do, providing that the seller has described the goods fairly and accurately. If the label on the dress says size 14, then I would say that the seller has fulfilled his or her part. However, if the label on the dress says 12, then of course you have the right to return it and ask for a full refund. See page 153 for resolving disputes on eBay.

There are exceptions to this. Professional sellers of clothes on eBay will allow you to return items providing you have not worn them. However, in nearly all cases you would have to pay for the postage cost to return the item to the seller.

Designer labels tend to be the best buys on eBay. You will find a good mix of new and nearly new clothes, together with a great number of fakes on eBay.

Useful websites

- **www.austinreed.co.uk** Men's clothes.
- **www.debenhams.com** Department store.
- **www.evans.co.uk** Women's fashion size 16 +.

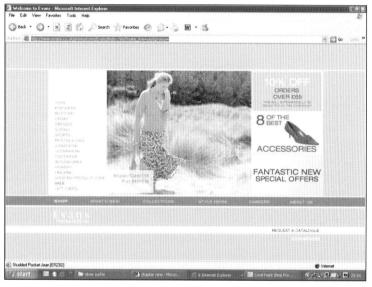

- **www.greatuniversal.com** Catalogue website.
- **www.lasenza.co.uk** Ladies' lingerie.
- **www.lauraashley.com** Ladies' clothes and household goods.
- **www.marksandspencer.co.uk** The high street retailer.
- **www.principles.co.uk** Ladies' fashions.
- **www.simplybe.co.uk** Ladies' fashions and homeware.
- **www.topshop.co.uk** Ladies' fashions.
- **www.ukshoppingcatalogues.co.uk** Browse and compare prices of womenswear at the leading mail order companies like Freemans, Kays, LX Direct, GUS, Littlewoods and more.
- **www.wallis-fashion.com** Ladies fashions.

Your reminder box

- Always check the size guide and the internet shop's return policy before placing your order.

- If a credit account has been set up, make sure you allow enough time for your payment to reach the shop and be processed.

- Don't shop on eBay until you have read pages 131–187.

Chapter 11

Health and Mobility Products

As we get older our need for health care increases. Although I believe the genes you have been given play a huge part in the ageing process, there is still lots you can do yourself to keep healthy and active – and the internet can help, of course.

Keeping active

A good workout in a gym is great for some people, but for any exercise programme to work it has to become a habit and not a chore. I didn't take to going to the gym very well, but I do enjoy a morning swim. So I cancelled my membership and took to going to the local swimming baths and saved a fortune! With the money I saved, I bought a motorised treadmill. For me, a treadmill is ideal; it gives me an aerobic workout that is good for my heart and I can watch the morning news at the same time. Exercising has become a habit, not something I dread or even have to think about. This is a really important part of keeping active: do things you enjoy! Just walking, if you are able, can really be of benefit, particularly if you make it a brisk walk. Keeping active and a healthy diet go hand in hand, so watch how many treats you eat, too.

Many of the goods and services you need to maintain a healthy lifestyle are available online, but before you start any exercise programme or splash out on a bike you've seen on eBay, it is important you see your doctor for advice.

Exercise equipment

Before purchasing any exercise equipment, you have to be sure it is right for you. Seek advice and try different equipment out before you purchase. There is always lots of used equipment for sale and often it has had very little use before its owner lost interest in it. This is one instance where it is perhaps best to take a look at the item you want in a high street shop so that you can try it out before shopping around online for a bargain price: and, of course, you won't have to carry it home.

Mobility aids

There is now a fantastic range of mobility aids available, from a simple walking stick to an all-terrain mobility scooter. Using mobility aids will help keep you active and independent. Before you purchase anything, seek good advice on the products that would suit your needs: never just ask a salesman!

Ask your doctor what help is out there, because some mobility aids are available on the NHS. Also, contact organisations such as Age Concern. As with all major purchases, you should research the product and, if you can, try it in a high street store.

Buying a mobility scooter

- Type **www.morethanmobility.co.uk** into your web browser.

- Now scroll down the range of different products in the left-hand menu and click **Scooters**.

- Now click **Portable scooter**.

- Now select any of the scooters and add it to your shopping basket. Select 1 for the quantity.

- You are now presented with a number of choices, including emptying your basket, updating it or going to the checkout, or you can continue shopping and add further products into your basket.

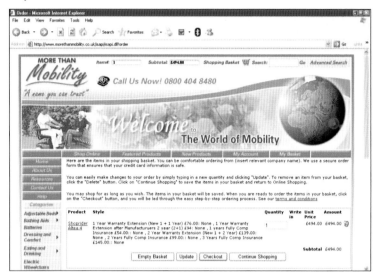

- For this exercise, select **Empty basket**, which will remove the products from your shopping basket. However, if you were really going to buy a scooter then you would select **Checkout**.

VAT exemption

You will not have to pay VAT on certain products if you declare that you are chronically sick or disabled. Chronically sick means that you have an illness that is likely to last a long time, such as arthritis or angina.

Disabled means substantially and permanently handicapped, by illness or injury. You do not have to be registered disabled to claim relief from VAT, but the nature of your illness or disablement must be specified. Nearly all websites have a online form for you to complete, as the government does require you to make a declaration. It is really just a matter of ticking a box and you will save 17.5%. There are no complex forms to complete.

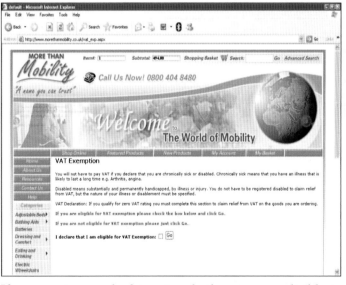

If you are in any doubt as to whether you are eligible to receive goods or services zero-rated for VAT, consult *Notice 701/7 VAT reliefs for disabled people* (see website **http://customs.hmrc.gov.uk**) or contact HM Customs and Excise Advice Service. Tel: 0845 010 9000.

Health products

As mentioned before, a healthy diet is really important to us all. If you are eating well and all the right foods, then you should not need any top-ups, but sometimes your diet may lack extra some essential vitamins or perhaps you have heard of a herbal remedy that has worked wonders for a friend. It is a good idea to speak to your doctor before you consider taking extra vitamins or herbal remedies, as many are stronger than you think and might interfere with prescription drugs. Also, too much of some vitamins can do you harm.

Private health care

Going private has always been thought of as expensive. Some of us consider private medical care when the NHS has let us down and when we are in need of urgent treatment. But if you think private health care is for you, it is best to take out a plan when you are fit and healthy, when the monthly payments will be relatively low.

It is never too late to join a private health care plan, but many will not cover any existing illness. As with most things in life, you get what you pay for. There are a number of plans to suit your pocket: the higher the premium you pay, the better the cover, but take your time and read the small print. See page 97 for more on insurance.

Unfortunately, we never know what lies around the corner when it comes to health. I have a friend who has paid into a health care scheme for years. Recently he had cause to contact them for help, only to be told that his level of cover was insufficient to cover the cost of his treatment. After many letters and phone calls, the provider did agree to cover half the cost, but he had to fund the remainder himself.

Useful websites

Exercise equipment

All these stores sell fitness and exercise equipment and supplies.

- www.argos.co.uk
- www.fitness-superstore.co.uk
- www.gymworld.co.uk
- www.powerhouse-fitness.co.uk
- www.reebokfitnessshop.com

- www.totallyfitness.co.uk

Mobility aids

- **www.activemobility.co.uk** Suppliers of disability and mobility equipment.

- **www.argos.co.uk** The online link to the superstore.

- **www.direct.gov.uk/DisabledPeople/MotoringAndTransport** Information on motoring and transport issues.

- **www.disability.gov.uk** General information on government services for the disabled.

- **www.mobilityextras.com** Mobility scooters and other aids.

- **www.mobilitymegastore.co.uk** Wheelchairs, mobility scooters and other aids.

- **www.mobilitypricewatch.co.uk** Scooters and other aids.

- **www.morethanmobility.co.uk** Supply and service of mobility aids.

- **www.motability.co.uk** Wheelchairs, scooters and car adaptions. They also offer short-break holidays.

Health products

These companies sell various health products.

- www.healthspan.co.uk
- www.healthydirect.co.uk
- www.hollandandbarrett.com

- www.magnetichealthcare.com

Private health care

All these companies offer private health insurance.

- www.bupa.co.uk
- www.privatehealth.co.uk
- www.standardlifehealthcare.co.uk

Your reminder box

- Keep as active as you can.

- See a doctor before taking extra vitamins or herbal remedies or starting an exercise programme.

- Find out what help is available on the NHS.

- You may be able to avoid paying the VAT on some disability products.

- If you are over 55, some health care plans may not be of any use. Check the level of cover provided.

- Always be truthful when completing health care forms, because in the event of a claim companies will look closely at your medical history.

Chapter 12

Holidays

Travel bookings via the internet have seen amazing growth in the last couple of years, with most hotels and airlines selling their services online. It is easy to see why more of us are booking our own holidays via the internet and making substantial savings. But before you rush off to Google to search for holiday deals, there is some really important information you should know to avoid your dream holiday becoming a nightmare.

Should I book an internet holiday?

If you have special requirements, maybe you need ground-floor accommodation for example, it is best only to book package holidays with well-known tour operators. They will make all the necessary arrangements for you, and if your hotel turns out to be unsuitable then they have a duty to find you one that is more appropriate, based on your requests made at the time of booking.

If you are reasonably active, by using a combination of tour operators' websites brochures to help research accommodation and other requirements, I am sure you will find a bargain holiday. Putting together your own holiday requires a little more research, but it can be really rewarding and can allow you to see more of the country you are visiting. The important thing is to take your time, and do all the research you feel is necessary in order to create the holiday you want.

Package holidays

Package holidays are booked with an established tour operator that provides you with a flight, transfers and accommodation. These tour operators are ABTA bonded and therefore, however you paid for your holiday, you have an excellent degree of protection. I have booked many of these types of holiday over the internet and have found some really good bargains over the years.

Incidentally, don't always take the insurance offered with your holiday. In most cases you can find a better level of cover if you shop on the internet. If you are a frequent traveller, consider an annual policy that covers you for a number of holidays throughout the year.

Non-package holidays

Non-package holidays are those you have put together yourself or someone other than a tour operator has put together for you. Suppose you book a cheap flight to France then find a bargain hotel in the centre of Paris. You pay for them both using your debit card. You've saved yourself a small fortune. What could go wrong?

Well, as you put the holiday together yourself, if the flight is cancelled you will just receive your money back. However, if you booked a package holiday with a tour operator, they would try to offer you an alternative flight or holiday.

Or you might find problems with the hotel. If you arrive to find it is unsuitable or dirty, there is little you can do other than to move room or hotel. Because you booked the hotel yourself you don't have a travel representative to complain to. You might speak some French, but perhaps not enough to explain the nature of your complaint to a non-English-speaker.

You can't rely on some of the internet travel agents. Some may mislead you into thinking you have booked a package holiday with a well-known tour operator by quoting leading names, when in fact it's only the airline that's well known. The accommodation may have no connection to the company that has provided the flight. You can usually detect this type of holiday by asking if the transfer to your hotel is included. The reply will be, 'It's only short taxi ride away.'

Don't worry **about going it alone**

If you book a non-package holiday using your credit card, you would be covered if, for example, the company folded and could offer no flights.

Self-drive holidays

If you own a car, then consider a short break in France or Belgium. It's closer than you think and I would recommend the Channel tunnel for your crossing. The main advantages are:

• The crossing is fast (around 35 minutes);

• You don't get sea sick!

• You simply drive into your carriage and the train takes you across;

• There are no steps to climb and you don't have to walk anywhere – you don't have to leave the comfort of your car unless you wish to stretch your legs;

• Making a booking is easy via the internet.

If you are going to take the car abroad make sure you have European breakdown cover, as well as the extra items you are required to carry in your car by most European countries – for example, bulbs, warning triangle, copy of your insurance, headlight beam deflectors. Check the requirements of the country you are visiting **www.theaa.com**.

Booking a trip through the Channel tunnel

- Type **www.eurotunnel.com** into your web browser.

- Enter the dates you are thinking of travelling and select the available crossing times. I visit the site often as it regularly has special offers.

Booking flights

There are many companies offering cheap flights to a variety of worldwide destinations. As with the holiday companies, it's important you consider the ones that are ABTA bonded. Nearly all airlines allow you to book direct with them and many have special offers throughout the year.

- Type **www.virgin-atlantic.com** into your web browser.

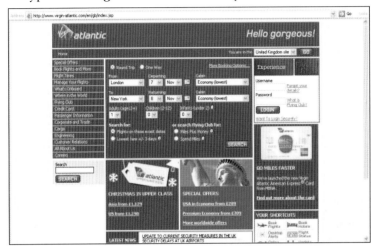

On this page, you are now required to fill in your date of travel, starting point, destination and the class of travel you require. Nearly all airline websites work the same way.

To continue to make a booking, you would need to complete the following steps:

- **Search:** This is the process of looking for flights that match your requirements. If you are prepared to travel two or three days either side of your preferred date, then you may find cheaper flights.

- **Select your flights:** Following a search, you will be offered a choice of available flights. Select the one that best matches your requirements.

- **Price:** At this point, the cost is usually shown.

- **Passenger details:** You are required to enter your details and those of your passengers. You must take your time over this and make sure all the details are correct, as any mistakes may cost you your flight. Make sure all the titles and names match the passports.

- **Payment:** You must now enter your credit/debit card details.

- **Confirmation:** When your confirmation is displayed, it is a good idea to note any booking reference details and, if possible, to print the page.

Some airlines also require you to register your details with them and will provide you with a user name and password. Make a note of these as they are used to log into the airline's website, should you need to change any bookings you have made.

Airport parking

There are many companies offering airport parking. When making a booking, check how far away from your airport terminal you will have to park, whether the car park is open or undercover and what security the car park offers.

Most airport hotels offer a 'park and fly' rate for people who have to travel a distance to the airport and have an early-morning flight. These can be a really good deal and are well worth exploring.

Useful websites

Disabled travellers

These sites offer specific travel advice for disabled travellers.

- www.abletogo.com
- www.disabilitytravel.com
- www.disabledholidaydirectory.co.uk
- www.tripscope.org.uk

Package holidays

These are among many sites for package holiday bookings.

- www.cruiseline.co.uk
- www.driveline.co.uk
- www.firstchoice.co.uk
- www.mytravel.co.uk
- www.packageholidays.co.uk
- www.thomascook.com
- www.thomson.co.uk

Booking flights

Shop around for flight bookings at recommended sites.

- www.airfrance.co.uk
- www.americanairlines.co.uk
- www.bmibaby.com
- www.britishairways.co.uk
- www.cheapflights.co.uk
- www.easyjet.com

- www.emirates.com
- www.firstchoice.co.uk/flights
- www.flymonarch.com
- www.mytravel.co.uk
- www.thomascookairlines.co.uk
- www.thomsonfly.com
- www.virgin-atlantic.com

Airport parking

Check airport parking options at any of these sites.

- www.compare-airport-parking.co.uk
- www.flypark.co.uk
- www.holidayextras.co.uk
- www.parkbcp.co.uk
- www.parking4less.co.uk
- www.PurpleParking.com
- www.SimplyParking.co.uk

Your reminder box

- Always try to make your bookings using a credit card.
- Consider finding a package holiday for your first internet-booked holiday. Do not try to put together your own holidays unless you are confident that you know what you are doing and understand the pitfalls.
- When looking for cheap flights, be prepared to move your date of travel by a few days as this can make a big difference when looking for bargains.
- Visit the websites of major airlines as often as you can. Many have special deals that are only available on the internet.

Chapter 13

Auction Sites

If you enjoy the search for a bargain, combined with the excitement of a real auction, then an internet auction site is something you should try. There are many available but eBay is probably the biggest and the best known. Remember: just as with a conventional auction, before using any auction sites you need to be sure of the rules and regulations surrounding the bidding process. In this chapter and the next, I will just look at buying on one of the biggest auction sites, eBay. If you wish to learn how to sell items and acquire a deeper knowledge of eBay, I would recommend you purchase a copy of *The Secrets of Successful Buying and Selling on eBay* published by Foulsham, available from all good bookshops and online from www.foulsham.com.

Don't worry **about auction sites**

If you don't feel confident about using eBay or any other auction sites, just ignore this part of the book. Once you get the hang of online shopping you can return to this subject.

eBay auction site

There can be few people who have not at least heard of eBay. It is a modern phenomenon, one of the fastest-growing businesses in the world, with an astonishing client base.

Like many of the world's largest companies, eBay has its roots in the most humble beginnings. In September 1995, Pierre Omidyar started eBay in his living room with an idea to create a community in which individuals and merchants could have equal opportunity to buy and sell new and used goods at fair prices – or in other words for his fiancée Pam to buy and meet collectors of the Pez™ sweet candy dispensers!

eBay has come a long way since then, with well over 100 million registered users buying and selling everything from Concorde parts and antiques to classic cars and the latest gadgets. The most expensive item sold so far is a jet for £2.8 million!

At first glance, the eBay website might appear complex, but don't let this put you off. It has been well designed, so it is really quite easy to use. It is also very sophisticated, with multi-layered elements and options. Within the next two chapters I will take you through the process of registering and buying on eBay. You'll soon discover the short cuts to success, as well as how to avoid the pitfalls.

How eBay works

Quite simply, eBay brings buyers and sellers together via its website. All sellers pay a listing fee and a commission to eBay when an item is sold. eBay makes no charge to buyers.

The online auction

An auction room works with a hall full of customers bidding on an entered lot; eBay works in just the same way but via the internet, thus creating a virtual hall where many millions of people can enter and bid.

If you want to buy, you simply turn up and bid. All items on eBay display the time and date when the auction will finish, and buyers can place a bid at any time before the item ends. In addition, some have a 'Buy It Now' price, which means you can purchase the item immediately for the price shown.

Getting started

Type **www.ebay.co.uk** into your web browser and it will take you to the UK website. If you live outside the UK, then type in eBay's web address for your country. For example, you would type **www.ebay.com** for the US.

The eBay home page

This will open eBay's home page. Since it changes all the time, it will not look exactly like the pages shown here, but it will be very similar. It will change again once you are registered. The key elements are always shown at the top menu on the home page, including 'Search', 'My eBay', 'Buy' and 'Sell'.

Becoming an eBay member

To buy or sell on eBay you must first become a member. It is quick and easy and is done via the eBay website.

* Type **www.ebay.co.uk** (for the UK) into your web browser. Your computer will open the current eBay home page. You'll see something similar to this screen:

- Click **Register**. This link will take you through the registration form.

- Fill out your name, address and contact details by clicking on each box in turn. Don't worry if you make a mistake. Any errors or missing information will be highlighted in red and you will be asked to fill out the box again.

Your e-mail address

Your e-mail address is how eBay and other members will contact you. It is important that you use an address you have regular access to. Most e-mail accounts allow you to access them from any computer, which can be useful for keeping an eye on your accounts while at work or away on holiday.

If you want to set up an e-mail address just for eBay use, you can do so free of charge at many websites: for example, www.msn.com and www.yahoo.com. However, if you use either of these you will be asked for your credit or debit card information. Don't worry: you will never be charged without your consent and your card information is kept strictly confidential. This process helps eBay to confirm who you are for its security purposes.

User agreement and privacy policy

During the registration process, you will find a link to details of eBay's terms and conditions. You should read them to ensure you are happy with them, then click on the three boxes confirming that you have read them and that you are 18 years of age or over.

Choosing your user ID

The name by which you will be known on eBay is called your user ID. Everyone chooses a user ID and a password so that only they can have access to their eBay information.

As your user ID is how other members identify you, you should spend a little time choosing something suitable. Try to think of an ID that is easy to remember, not too complicated, and perhaps describes your interests. It is best to avoid obscene or silly names; they might seem funny at the time and appeal to you, but many members might be put off when buying from the 'killerkid' or 'dodgydan'!

To make choosing a user ID easier, eBay will suggest a few user IDs, usually linked around your name. If you are happy with one of these, click next to it. Otherwise, fill in the name you want.

- Once you are happy with the name you've chosen, click **Create your own ID**.

You can change your user ID at any time, although if you do this, a 'Changed ID' icon is shown next to your ID. This can deter sellers (and buyers) if they suspect this may be an attempt to hide your eBay past.

User ID already taken

With the many millions of eBay members, there is a good chance your first choice of user ID may already be in use. If this happens, eBay will guide you through choosing a different ID and will even offer you some suggestions.

Your eBay password

Your eBay password is the only means you have of preventing others from having access to your eBay account. When signing into your account, you will be asked for your user ID and your password. It is really important you tell no one of your password and make sure you change it regularly.

I recommend that your password should:

- Be six to eight characters long;

- Be made up of a mixture of capital letters, numbers and special characters;

- Not include words that can be found in a dictionary;

- Be impossible to guess.

If you submit a poor password – your name, for example – eBay will reject it and ask you for another.

Security meter

To help you with password security, eBay has come up with a security meter. As you type in your chosen password, a bar appears; the darker the bar, the more secure the password.

Secret question

If you ever forget you password, eBay will ask you to confirm your answer to your selected secret question before sending your password to you by e-mail. It provides a list of options for you to choose from: for example, 'What street did you grow up on?' or 'What is your mother's maiden name?' You must select a question and type in your answer, making sure that it is a reply that is easy to remember.

Finalising your registration

You will be pleased to know that you are almost there with regards to registering an eBay account and becoming a member.

Confirming your identity

eBay needs assurance that you are who you say you are.

If you have used a Hotmail or a Yahoo e-mail address, you will be asked to confirm your identity by providing debit or credit card details.

 Don't worry about confirming card details

If your credit card details are requested as part of the process of confirming your identity, this is only an extra security check. You will not be charged and your card details are kept safe within eBay.

If you have used an e-mail address offered by an internet service provider such as AOL or Wannado, you will not be required to enter your credit card details.

Check your e-mail

You will then be taken to the 'Check Your E-mail' web page, either direct from the previous page if you are with a standard ISP, or once your card details have been accepted if you are with Yahoo or a similar provider.

Open the e-mail account that you provided eBay with during registration and look for the e-mail it has just sent you. If nothing appears in your inbox, check your junk mail folders, as some ISPs or spam filters will reject this kind of e-mail. In this case, identify the sender as 'Not spam' and retrieve the e-mail. This is the only time you need to click on a link in an e-mail from eBay. Fake e-mails are circulated in which you are asked to confirm your name, password and/or credit card details. They will look authentic; they are not.

Once you have opened the e-mail, click on the link. This will confirm your registration.

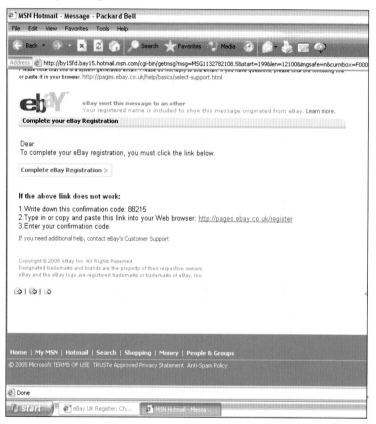

Signing in and out

That's it! Now you are an eBay member. Each time you visit the site, you will be asked to sign in, using your user ID and password.

- To sign in, go to the eBay home page, then click **Sign in** at the top of the page.

- You will then be asked for your eBay user ID and password. Fill out the boxes by clicking on each one in turn, then click **Sign in**.

You can then go wherever you like on the site, using the menus and links provided.

When you have finished visiting eBay, always sign out to prevent anyone else from using your account. This is especially important if you have visited eBay from your work station or an internet café.

• From the home page, simply click **Sign out**.

When you have successfully signed out from your account, you will see a message to confirm that you have done so.

My eBay

Each time you sign in to your account, you will be taken to the eBay home page. It will look slightly different every time you visit because it is constantly updated, but the top menu will remain roughly the same. You should see the message 'Welcome to eBay', then your username.

• Now click **My eBay**. This is one of eBay's most useful features and where you should spend some time exploring because it contains all the information connected with your account: items you are bidding on, things you have bought (and sold), feedback, messages and so on.

My eBay site map flowchart

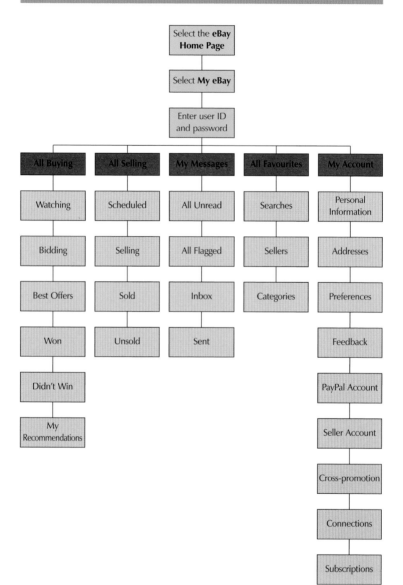

'My eBay' is full of so much information that it may seem a little complicated at first glance. But if you take it step by step, investigating sections as you need to use them, you will soon find your way around.

The 'My Summary' section shows your recent account activity, listing the details of any items you are buying or selling. To start with, this is the most useful section. It will show you the details of the item, whether you are the highest bidder, how long before the auction ends, and other useful information. This section also includes 'Items I'm Watching'. If you do not want to bid on an item but would like to keep an eye on how the auction progresses, you can add it to this list.

On the left side of the screen you will see a list of links associated with your eBay account. Here are just a few:

All buying

- **Watching**: Lists the items within your watch list;
- **Bidding**: Displays items you are bidding on;
- **Best offers**: Shows items that you have made a best offer on;
- **Won**: Lists 'Buy it now' items and items you have won at auction;
- **Didn't win**: Lists items you did not win;
- **My recommendations**: Lists your recommended sellers.

My messages

Displays your messages from eBay. In most cases you will also be sent a copy via your e-mail address.

All favourites

- **Searches**: Useful feature that lists your favourite searches;
- **Sellers**: Shows your favourite sellers;
- **Categories**: Lists your favourite categories.

This has just given you the basics and scratched the surface of 'My eBay', so when you have some spare time, look and click on the many links and discover for yourself what they all do.

Searching eBay

With the millions of items offered on eBay, it might feel as though finding what you are looking for will take you hours, if not days, but this is not so. eBay's search tools are really powerful and will take you to what you are looking for in seconds (or slightly longer if you are not on broadband!). You will find that eBay's search box works in just the same way as internet search sites.

The simplest way to find items offered for sale is to use eBay's search box. This can be found on eBay's home page and within 'My eBay'. As an example, search for a piano.

- Type **piano** into eBay's search box. At the time of writing, 'piano' produced over 3,500 items offered for sale.

If you had plenty of time to kill, you could look at each individual item; good luck! But the search can be narrowed down if you are more specific. Let us assume that you are looking for an electric piano.

- Type **electric piano** in the search box. At the time of writing, 'electric piano' produced only 18 items, a much more manageable number.

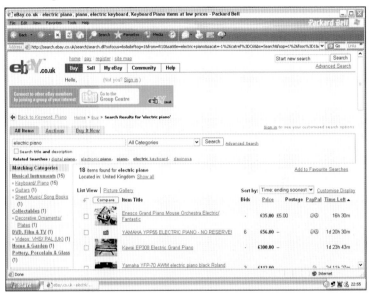

You can also narrow the results by searching for a particular item within a specific category; for example, type **piano** in the search box, then use the drop-down categories menu to search only within musical instruments. The only downside of this option is that not all sellers will list their items in the appropriate categories, or you may be looking for an item that could cross into a few categories. Always try to be specific in the keywords you use for your search, and think about other words that sellers may use to describe their items.

For rare items, tick the **In titles and description** option, or click **Advanced search** and fill in the boxes with more details.

As you become more experienced, you will soon discover the best way of using and combining keywords to find the items you are looking for. Some of the best bargains I have found are the ones that have been misspelled by the seller, so other people have not found them on their searches and there have been fewer – or no – bids! Searching for items under one word or two will also make a difference to the results: for example, you will get a different set of results if you search for 'Bladerunner' than if you search for 'Blade Runner'.

Saving your favourite searches

Once you have found a way of searching using keywords, categories and so on, eBay will remember it for you and will even send you an e-mail when new items are offered in your favourite categories.

* First, enter your favourite search information, then click **Search**. On the top left-hand corner of the results page, click **Add to favourite searches**.

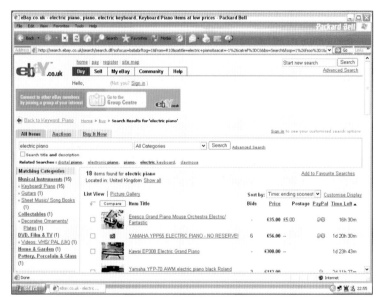

- If you have not already done so, you will be asked to sign in to your account. On the 'Favourite searches' page you will see an e-mail tick box. If you do not want an e-mail informing you of new items offered that match your search criteria, untick the box.

Your favourite searches can be viewed at any time and can be found in 'My eBay'.

eBay payments

Following a successful bid on eBay, you will have to deal with sending a payment. There are a number of payment methods that can be used, depending on the requirements of the seller. When you pay for an item, the seller will usually indicate a preferred payment method.

PayPal

PayPal is my preferred method of payment, and that of most sellers on eBay. It is by far the safest and quickest method to use and allows you to send and receive money swiftly and securely, nationally and internationally. For more details, see pages 48–64.

Cash

Cash is one of the oldest methods of payment and I have used it for collecting goods in person. However, I would never recommend sending cash, even small amounts, by post (even though most sellers would accept it, providing it is sent at the buyer's own risk). It is simply not safe or reliable and is impossible to track.

Cheques and postal orders

For UK transactions, cheques and postal orders are easy and relatively safe. The only disadvantage of a cheque is that most sellers will not send items out until your cheque has cleared, so you will have to wait a few days for your purchase.

Money orders

Money orders are available from most banks. Used internationally, they are converted into the local currency in cash by the recipient, so must therefore be treated in the same way as cash. My advice is never to trade with buyers or sellers who ask for money orders. Criminals use money orders because they are hard to track and give instant cash. Also – and especially if you are buying any high-priced item – take some time to check out the seller's history and credentials before you make a bid.

Bank transfer

A bank transfer moves funds directly to or from your bank account. I am never happy to give out my bank details and would never use this method for high-value items or international purchases.

Resolving payment disputes

Mistakes do happen, so if you ever have a problem with a purchase (or sale), contact the seller (or buyer) first. You can do this from the drop-down menu next to the item on 'My eBay'. Remember: co-operation is better than confrontation. Only when you have done this and you find yourself making no progress should you use PayPal's resolution centre.

- Log in to your PayPal account, then select **Resolution centre**.

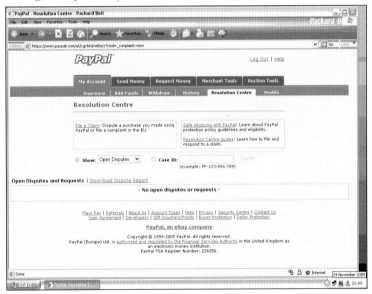

- Simply follow the straightforward procedure to log your claim. You will receive an e-mail acknowledging your dispute information, followed by e-mails to let you know what has been done and how the dispute has been resolved.

Protecting your account against fraud

Protecting your eBay and PayPal accounts is really quite simple:

- Change your passwords regularly;

- Never reveal your password to anyone;

- Check your accounts regularly for any suspicious activity;

- Do not click on any 'Sign in' links in e-mails.

Fake e-mails

From the day you open your account with either eBay or PayPal, you may be targeted by both amateur and highly professional crooks. They are hoping you will fall for their scams and fake e-mails and reveal your passwords and bank details, thus allowing them to steal from you and use your identity for bigger frauds.

Some fake e-mails are very easy to spot as they are written in very poor English, but the majority are convincing and well presented, copying eBay's logo and layout and showing the sender ID as eBay or PayPal (see page 26).

If there is one thing you remember from this chapter, then please let it be this: never reply to any e-mail asking you for your eBay or PayPal user ID, passwords or bank details, even if it looks like it is from eBay or PayPal.

All these fake e-mails have one thing in common: they ask for your password or bank details. To encourage you to reveal this, most e-mails will request you verify your account details as a matter of urgency to prevent your account being restricted or closed. Some will ask you to click on a link; this link takes you to a fake 'Sign on' page identical to eBay's own. The fraudsters use this fake page to capture your details.

If you receive any e-mail asking for confidential information, do not reply to it. You can report it via e-mail to **spoof@ebay.co.uk**.

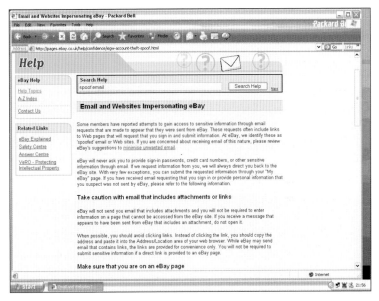

Another security measure is always to go direct to the eBay and PayPal pages. I would recommend you always open a new web browser when you access either site and type www.ebay.co.uk and for PayPal use www.paypal.co.uk when you wish to sign in to your account. Log in to your accounts regularly to check for any suspicious activity.

Imaginary items

As an eBay member, you may occasionally come across items for sale that don't exist. For example, I found the latest plasma television with a starting bid of 99p. I placed the item within my watch list and waited to place a bid towards the end of the auction. However, when I read the description carefully, I realised that I would have been bidding on an e-mail that will be sent to the highest bidder explaining where one could purchase the item at a 50% discount.

I have never heard of anyone who has purchased items this way or who has saved money doing so. It is always a con to attract lots of people to part with a small amount of cash each.

There are many spins to this. Some say you will receive a call giving you the information, others a voucher or a link to a special website where the goods are on sale. They all lead to nothing. Stay clear of them and look for genuine bargains.

The long fraud

This is one of the most effective types of fraud found on eBay. A seller opens an eBay account and waits for, say, a year, building up positive feedback by purchasing small-value items from genuine sellers or selling the items to him or herself, using a different user ID. This is called shill bidding and is against eBay rules.

Then the seller will offer a high-value item such as a plasma television for sale. The seller might also ask to be paid using a money order. The item never arrives.

Postage scam

Many eBay sellers will put a little extra on the true cost of postage and packing. This is very common as it gives sellers something for their time in packing up the parcel and taking it to the post office, and covers their seller's fees if the item sells at a very low price. However, there are some sellers who will make more on postage than they do for selling the item itself!

Another spin to this scam is for the seller to hide the actual price of an item in high postal charges. For example, I was looking for a notebook computer on eBay and found one that was new and had just 30 minutes left to run, with a bid of only £55.00. I thought it strange, so I looked carefully at the description.

The seller was charging £700 postage and packing, thus avoiding paying eBay's full fees on this expensive item. This might not seem a bad idea, but the problem occurs when things go wrong and you enter into a dispute with the seller. Instead of a dispute involving a £755 notebook computer, in the eyes of eBay it becomes a £55 notebook computer.

Always check the postage costs carefully and make sure you feel they are fair before you bid. Also make sure that the seller is not going to add VAT, as some VAT-registered users will try this. It is usually written somewhere in the item description. If a seller ever tries to add VAT later, refuse to pay it.

Useful websites

As well as eBay, there are a number of other auction sites.

- **www.bid-up.tv**
- **www.christies.co.uk**
- **www.eBay.co.uk**
- **www.ebid.co.uk**
- **www.sotherbys.com**

Your reminder box

- Choose your password carefully. Never write it down or share it with anyone, and change it regularly.

- If you have any doubt that an eBay-related e-mail is genuine, do not open it. Never reply to any e-mail asking you for your eBay or PayPal user ID, passwords or bank details, even if it looks like it is from eBay or PayPal.

- Always read item descriptions carefully before you bid, so you know exactly what you are bidding for.

- Never send cash through the post. You cannot guarantee that it will arrive, and there is only your word that you sent the right amount and the recipient's word that it did or did not arrive.

- Beware of sellers who insist on money orders.

- If you purchased something using a credit card, you may also contact your card issuer as it can offer you additional protection.

Chapter 14

Buying on eBay

Now you understand the basics, you are probably itching to get going and buy something on eBay. If you just take a little time to run through this chapter, then you will find it easier and you are more likely to be successful.

The bidding sequence

You will understand the information in this chapter better if you know the basic sequence given below, but it is best to read through the whole chapter before you proceed.

- Sign in to your eBay account.

- Search for items you want using the search engine.

- Click on the item to take you to the product page.

- When you are sure you want to place a bid, click **Place a bid**.

- Fill in the maximum amount you want to bid for the item and click **Continue**.

- Check the details shown on the next screen and click **Confirm bid**.

- If you are the highest bidder, this will be confirmed. If not, you can raise your maximum bid.

- You will receive an e-mail confirmation and the item will be listed in 'My eBay' under 'Items I'm bidding on'.

- If someone outbids you, you will receive an e-mail to let you know and encourage you to bid again; the item in 'My eBay' will appear in red instead of green.

The path to success

There are a few simple keys to successful buying on eBay.

- **Research the item:** Take your time to read the description carefully and make sure it is what you want. Look at the picture, check the specification and read the small print.

- **Check anything you are not sure about:** You can send an e-mail to the seller via the product page.

- **Research the seller:** Check the seller's feedback to make sure he or she is a reliable source.

Understand that popular items such as plasma or LCD televisions, MP3 players and computer games will attract lots of interest, as these products are always in demand. Therefore, you are unlikely to make huge savings when purchasing these. The real bargains are on the unusual and specialist items. If you are lucky and have followed my search advice you might have found items that are listed incorrectly or misspelt. These attract less interest and give you an opportunity to purchase a real bargain.

Buying on eBay flowchart

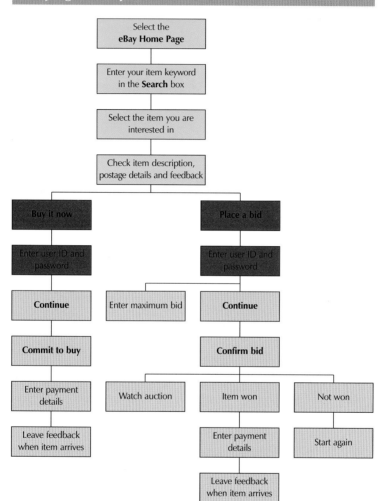

The winner's curse

'It's only worth what someone is willing to pay' is a cliché my father was fond of. When a seller lists an item on eBay it is offered to millions of potential buyers. It would be fair to say that the final price the item achieves in most cases reflects the market price. Therefore, if you are the highest bidder and have outbid everyone else, you have paid more than your rivals considered the item was worth. The more rivals watching and bidding, the less chance your item was a real bargain. This line of thought is generally forgotten, especially in the excitement felt during the closing few seconds of an auction. Before you start to bid, think about the 'winner's curse' that states that the act of wining rules out getting a bargain because you have paid more than someone else was willing to pay.

Don't worry **if you miss out on a bargain**

Console yourself with the fact that whoever won the bid might have paid over the odds for it and, anyway, eBay is full of great deals.

What not to buy

Amongst the many millions of items for sale on eBay, a number of items are prohibited. Although eBay does its best to prevent many of these items from being listed, some will always get through. If you do attempt to purchase any such items, you do so at your own risk as prohibited items fall outside any buyer protection offered by eBay or PayPal. A full list can be found in eBay's 'Help' section under 'Prohibited, questionable and infringing items'.

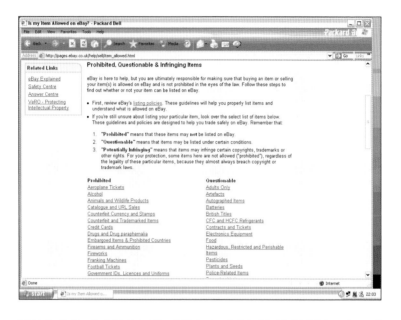

Who not to buy from

Most sellers on eBay are honest and reputable, but as in anything, there are a few bad apples and you need to be able to spot them. You can do this using eBay's feedback system. This is similar to the comments or visitors' book you might find in a hotel, but in an electronic form. Following a sale or purchase, both parties are invited to leave comments. These comments are listed on the member's site and can be seen by all eBay members.

Always check the feedback score of the seller. Look at the positive feedback and when the person became a member. In general, I would trust someone who has lots of positive feedback and who has been a member for over a year. As an added protection, never send money orders or cash. Instead, use PayPal and your credit card, especially for larger items.

If you do bid successfully for something that has been incorrectly described in an attempt to attract bids, don't pay for it but contact the seller. If you have already paid, insist on a refund.

Checking feedback

Take the trouble to check a member's feedback. It is easy to do and could well stop you wasting your time and money.

- Click the user ID shown within the seller information box found on the right of the auction page.

You have just selected this member's profile.

Now study the following:

- **Feedback score:** The feedback score represents the number of eBay members that are satisfied doing business with a particular member. It is usually the difference between the number of members who left a positive rating and the number of members who left a negative rating.

- **Positive feedback:** This represents positive ratings left by members as a percentage.

- **Members who left positive:** This represents the number of members who have left a positive rating.

- **Members who left negative:** This represents the number of members who have left a negative rating.

- **All positive feedback received:** This represents the total number of instances of positive feedback received for all transactions, including repeat customers.

- **Recent ratings:** This shows all of the ratings left for this member during the past month, six months and 12 months.

- **Bid retractions**: This shows the number of times a bid has been retracted (see page 179).

In an ideal world you should buy from a member who has a high feedback score together with high positive feedback. Click on some of the transactions to see if they are for small amounts or large amounts, and read the feedback that has been left. If you have any doubts, do not buy from them. With the many millions of items appearing on eBay, it won't be long before it appears again, this time offered by a seller you can trust.

Choosing your item

Once you have found the item you are interested in, you need to consider the following:

- Do you really want this?
- What is the cheapest price new in the shops?
- Is the item new or used; if used, how used?
- Is there any warranty?
- Have you read all of the item description?
- Is the description accurate or vague?
- Does the picture match the item or is it a sample?
- How much is postage?
- Is the feedback good?

Knowing the seller

If you are going to bid on expensive items, try to get to know your seller by asking questions about the item. You can learn a lot from the replies and the speed of the response. You should always ask the seller questions about the item you are about to bid on. Start by asking questions that will confirm the condition and elicit any information you would like that is missing from the item description. Be specific. If you don't get a reply, this could mean that there is a problem with the item or the seller, so I would not proceed any further.

Do your research. If a seller suddenly switches from selling cheap items to expensive luxury goods, alarm bells should ring.

How to ask a question

- Access the auction page of the item you are interested in. Then select **Ask seller a question** found within the seller information box.

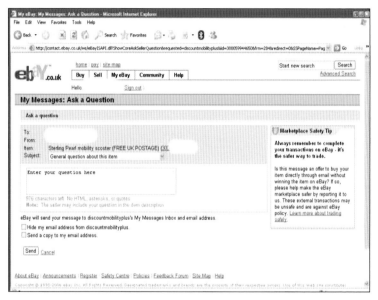

- Click on the type of question you wish to ask from the drop-down menu.

- Next type in your question and select **Submit question**. Before your question is sent you may be asked to copy the characters from a random text box into the blank box. This is to prevent automated questions being sent.

If the seller answers your question, then a copy of the reply is sent to your e-mail address.

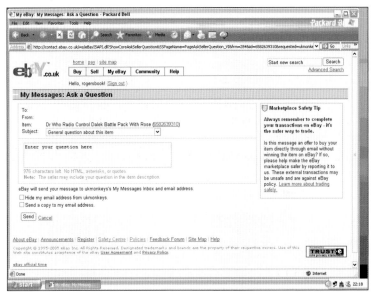

Don't be afraid to ask questions of the seller. If the answers aren't forthcoming, or if you are not happy with the replies, you could save yourself an expensive mistake.

The reserve price

Just like a non-internet auction, eBay offers its sellers the opportunity to set a reserve price. Buyers do not know the reserve price until bidding has exceeded it. Then a 'Reserve met' message is displayed within the listing. eBay sets a £50 minimum starting reserve, so all reserves are for £50 or more on all listings.

Checking postage and packing

Before placing a bid, check the cost of postage and packing, and whether the seller will post the item to you. Most sellers will act honestly but some try to increase their profits by charging a much higher cost than you should be paying. If this happens to you, ask the seller, 'Why so high?' The chances are you won't get a reply, in which case I would not proceed with placing a bid.

Watching items

With all the items ending at different times you might think you will need to write them down. With your very own 'Watch list' you don't have to.

• From the auction page of the item you are interested in, select **Watch this item** from the top right of the page. Once selected, it is saved to your 'My eBay' page without you having to make a bid.

• To view your 'Watch list', select **My eBay**; all the items are displayed within the summary.

- If you wish to remove items from your 'Watch list', simply click the box next to the item, which places a tick within the box. You can tick any number of items at once.

- Then select **Delete** and the item is removed.

Bid history

As you watch an auction it is worthwhile keeping an eye on the bid history.

- From an auction page, select **Bids**. This page shows you the full history.

As you become more experienced, you will be able to gain lots of information from this, including:

- Number of bids;

- Number of bidders;

- Type of bidders and their habits;

- Level of interest.

Comparing auctions

When searching for an item, it is likely you will be presented with a number of auctions as a result. You could make a note of each one, but there is an easier way. Comparing auctions can be useful when more than one of the same item is on offer.

• From your 'Search results' page, click the small box next to the item you wish to compare.

- Now click another item you wish to compare. When finished, just click **Compare**. You will then see the items displayed together.

Now you also have the option to add them to your 'Watch list' by clicking **Watch all** at the top of the page. The 'Compare' function will also work on all items within your watch list.

Buy it now

From time to time you will come across a listing with a 'Buy it now' option. This allows you to purchase the item instantly without the need to bid or wait until the auction finishes. There are two types: a fixed 'Buy it now', where the price displayed is the price you pay, just like in a shop; and an auction with a 'Buy it now' in which you are invited to bid for the item but have the option to end the auction early and purchase it at the 'Buy it now' price.

Best offer

New to eBay is an option called 'Best offer' located next to the 'Buy it now' option on the listing. If you see 'Submit your best offer', then you select this link to make the seller an offer. Both buyers and sellers should understand that this forms the binding terms of an agreement, if accepted. Once your offer has been sent, you may track the progress within 'My eBay'. The seller has 48 hours to think about your offer.

If the seller turns down your offer, the word 'Declined' will appear. If your offer is accepted, then you will be sent the 'Winner' e-mail and you must complete the sale as usual.

When to make a bid

Resist the temptation to rush in and bid, unless there are only a few seconds left before the auction ends. Add the item to your 'Watch list' so it can be found again quickly and easily within 'My eBay'.

You may place a bid any time before the auction ends. The best time to make a bid is during the closing few seconds of the auction. This is known as sniping. If you can't be at a computer during the closing few minutes of the auction – and remember you can use any computer, it does not have to be your own – then you must make your bid as late as possible.

Sniping for beginners

There are many bidding strategies, all claiming to improve your chances of winning the auction. Most experienced eBayers understand how the bidding system works and will formulate their own system, which I am sure in time you will do. For now, I will tell you one of mine:

- First decide your maximum bid, taking into account the condition of the item and the postage cost. Stick to this price and never increase it.

- Place two maximum bids, one during the last two minutes and the other during the last 30 seconds or less, depending on the speed of your internet connection.

- Bid a round amount first, followed by one with a few more pence added. This is because most eBay members bid in round numbers. For example, let's say the maximum you wish to bid is £10 and a few pence. First place a maximum bid of £7 two minutes before the auction ends. Now place another maximum bid of £10.67 regardless of whether you are winning or not. The eBay system will always take the higher amount, should another member bid against you.

Don't get caught up in a bidding frenzy. In the heat of the moment you might find yourself bidding – and paying – well above your planned maximum bid.

Making a bid

All being well, you will have done all your research, you have a maximum bid in your head and you are now ready to bid.

• Open the auction page you are interested in and click **Place a bid** and the bid window will open up.

• Enter your maximum bid. Remember: if you are sniping, this should be the first of your two maximum bids. If you can't be there to watch the end of the auction, place your maximum bid now and click **Continue**.

- In the next window, you will be asked to review and confirm your bid.

- You are about to make a bid that is legally binding. Check all the details, including your maximum bid, and only when you are happy should you click **Confirm Bid**.

Bid confirmation

- If you have been successful, the message 'You are the current high bidder' is displayed.

- If your bid was unsuccessful, the message 'You have been outbid by another bidder' is displayed. You will then be invited to bid again.

Checking your bid throughout the auction

It is important to check your bid status often, especially in the closing few seconds. During an auction, if you have been outbid, an e-mail will be sent to you, but this is not as quick as checking your own bid status. You can do this as follows:

- First open the auction page you are interested in.

- Click **Refresh** within your browser. This updates the page including the time remaining and bid information.

- During the closing few seconds of an auction you may find yourself refreshing the page many times. For many members, this is the most exciting part of eBay.

- If you have been outbid, you may place another bid anytime before the auction ends.

Bidding increments are set by eBay. If you want to check the increments, you can search its site.

Bid retraction

It is possible for you to retract your bid, but by bidding you have committed yourself to a legal contract. There are only a few circumstances in which you may retract a bid:

- You clearly entered the wrong bid amount – for example, £1,000 instead of £10.00;

- You can't contact the seller;

- The description does not match the item.

This is how you retract a bid:

- Open the auction page you have placed a bid on, then make a note of the item number. You will need this later.

- Next select **Bid history**.

- Click **Retract your bid**, located at the bottom of the 'Bid history' page.

- You must now read the 'Bid retraction' form. If you are happy, continue by selecting **Bid retraction form** at the bottom of the page.

- Enter the item number and select a reason for your bid retraction from the drop-down list. Finally, click **Retract bid**.

It is very important you don't do this too many times because bid retractions are recorded and can be seen by all eBay members. As an alternative, you can ask the seller to cancel your bid.

Item not won

If you have been outbid and have not won your item, don't worry. Any fool can win an auction by simply paying much more than the item is worth! Remember: there are new items appearing on eBay all the time so be patient and I'm sure you will have another opportunity to bid.

How to pay

If you are the highest bidder when the auction finishes, congratulations: you are a winner. A message 'The item is yours' will appear within the auction page and an e-mail will be sent to you.

Below the congratulations message on the auction page you will find a 'Pay now' link and the methods of payment accepted by the seller. You will also find this and payment information within an e-mail that will be sent.

Remember that all items that are won and that you are watching or bidding on can be found within 'My eBay'.

Select **Pay now** and choose your method of payment from the options preferred by the seller. Once you have completed this, you can check the status within 'My eBay'. If you are sending a cheque or cash, don't forget to include a copy of the auction page with your name and address. If you don't have a printer, write down the item number with a description and your name and address and enclose this with your payment.

Leaving feedback

By now you should have a good understanding of the feedback system and appreciate that it is at the heart of eBay (see page 163). It is what is used to judge other eBay members. Positive feedback helps members sell their goods, whereas negative feedback warns off buyers. Before you ever leave negative feedback, you must work with the seller or buyer to resolve any problems. Only when you can't resolve problems should you leave negative feedback.

As well as positive/negative feedback there is neutral feedback, but this isn't used very often because many eBay members regard it just the same as negative feedback. I would use neutral feedback if the seller were difficult to deal with or slow to help if there were problems with the item, but if after much time, trouble and persistence – rather than the seller wishing to help – it was sorted out and I was happy with the item.

Other eBay members will rely on your feedback, so it is important that you submit it promptly and accurately.

• To do this, click **My eBay** followed by **Feedback** located within 'My summary' on the left-hand menu.

- Now simply click **Leave feedback** next to the relevant item.

Nine times out of ten you will click Positive feedback. Then leave your comments – for instance, good to deal with, quick purchase, excellent. Remember to check the feedback you leave before selecting **Leave feedback**. Once left, it is virtually impossible to change.

Resolving problems

Item does not arrive

If after a while your item does not arrive, first contact the seller and ask for confirmation of when the item was sent. The seller might have simply forgotten to post it. Be friendly as the problem could be with the courier and not the seller. Always try your best to work problems out with the seller, but if you are making no progress then contact eBay.

You can report your problem and the seller to eBay through the 'Help' link. eBay will then begin investigating and, all being well, resolve your problems.

Damaged and poorly described items

Always contact the seller first in a timely manner. The quicker you report problems to the seller the better, because there may only be a limited time to make a claim with the courier. It is vital to keep all communications friendly and polite.

I once received a damaged toy. I suspected the seller knew this because he said he had 'never tried' the part that was damaged. The damage halved the value. The seller agreed and offered a 50% refund if I wished to keep the item. I accepted, as it was a bargain to begin with. Only after you have tried your best should you report the seller to eBay and start proceedings to recover your money for damaged or badly described items.

Seller disputes

If you have followed my advice and researched both the item and the seller, then disputes should be rare. However, there are some bad apples, and eBay has established a system for minimising their impact and looking into problems. If you have tried your best to communicate with the seller without success, you should report him or her to eBay. Select **Help** and then search **Dispute resolution** and follow the instructions you find there.

Your reminder box

- Have a good look around eBay and familiarise yourself with the site before you begin bidding on items.

- Set yourself a bidding limit and do not exceed it.

- Only bid on items that you are sure you want.

- Don't get carried away by a bidding frenzy as the end of the auction approaches.

- Exercise the same caution as you would when buying from any other source. Ask questions and if in any doubt, walk away.

- Keep an eye out for fake e-mails.

- Check out the seller's feedback and be prompt and honest when making your own.

- Consider post and packing charges when deciding whether or not to bid.

Index

accessibility 11, **13–66**
advanced searches 11
airlines 126–7
airport parking 127, 129
anti-virus software 18, **22–3**
Asda's website 82
auction sites 10, 130, 156
    see also eBay

backing-up files 24
bank transfers 152
bidders 11
bidding 174–6
    sequence 157
    shill 155
    strategy 173
bids
    checking during auctions 178
    confirmation 177
    history 168
    retraction 178–81
books 93, 100
broadband routers and
    firewalls 18
browsers 11
buying on eBay
    buy it now 171
    clothes 108
    flowchart 159

car insurance 124
cash payment 151
CDs, saving data onto 24

Channel Tunnel 125
chargeback 45
checkout 11
cheques 45, 151
clothes 103–8
credit accounts 103, 110
    and eBay 108
    ordering online 105–8
    returns policies 104
    size guides 103–4
    websites 109
computers 90–2, 98
consumer protection 44–5
credit accounts 103, 110
credit cards 44–5, 46–7
    and consumer protection
        44–5, 63, 124
    and eBay 137
    and PayPal 55

debit cards 44, 46–7
    and PayPal 55
delivery
    catchment areas 65, 71
    costs 38, 40, 85
    groceries 65
disabled
    mobility products 112–6
    travellers 122, 128
    VAT exemption 116
    see also accessibility
dishwashers 88

e-mails
  addresses and eBay 133
  fake 12, **26–8**, 153–4
  preventing viruses 17–8
eBay 45, 130–2
  bargains 148–9, 158, 160
  becoming a member 132–9
  best offer 172
  bid confirmation 177
  bid history 168
  bid retraction 178–81
  bidding 174–6
  bidding sequence 157
  bidding strategy 173
  buy it now 171
  buying clothes on 108
  buying flowchart 159
  checking bids during
    auctions 178
  comparing auctions 169–71
  fake e-mails 153–4
  favourite searches 149–50
  feedback system 161–4,
    184–5
  finalising registration
    137–9
  fraud protection 153–6
  imaginary items 154–5
  My eBay 142–5
  passwords 135–6
  payment disputes 152, 186
  payment 151–2, 182–3
  postage scams 155–6, 167
  problems with items 185–6
  prohibited items 160
  questioning sellers 164–6
  registration 132–9
  research 158, 161–6
  reserve prices 166
  returns policy 108
  searching 146–50
  secret questions 136
  signing in and out 140–2
  sniping 173–4
  user IDs 134–5
  watching items 167–8
  see also PayPal
electrical goods 85–91, 97
electricity 94–5
energy 94–5, 100–1
  saving appliances 87, 88,
    89, 97
exercise 111
  equipment 112, 118
express shopper 74

favourites 73
  on eBay 149–50
firewalls 11, **18–21**, 22–3, 25
  buying 22–3
  switching on 20
  updating 21
flights, booking 126–7,
  128–9
fraud 26–30
  eBay 153–6
fridge freezers 89
fuel 94–5, 100–1

gardening products 92, 99
gas 94–5, 101
Google 32–5
groceries 65–6
  delivery 65

express shopper 74
offers and deals 76
shopping websites 71–80
see also supermarkets

hardware 11
and firewalls 18
health care, private 117, 120
health products 111–2, 117,
118, 120
holidays 122–9
booking flights 126–7,
128–9
disabled travellers 122, 128
insurance 123
non-package 123–4
package 123, 128
self-drive 124
hyperlinks 11

icons 11
installation costs for new
products 38, 40, 86
insurance services 96, 102
car insurance 124
holiday insurance 123

keyboard shortcuts 15
keywords, choosing 33–5

laptops 90, 91
logging in and out 71
eBay 140–2

menus 11
mobility products 112–6,
119

scooters 113–5
VAT exemption 116
money orders 152

Next 105

offers and deals 76
Outlook Express 17–8

package holidays 123, 128
padlock symbol 46, 47
passwords 29
eBay 135–6
paying for goods 44–7
eBay 151–2, 182–3, 186
groceries 80–1
payment disputes 63, 152,
186
and security 46–7
PayPal 11, **45–6**, 152
adding credit/debit
cards 55
adding funds from a
buyer 59
adding funds from your
bank account 60
fees 52
paying through 61–2
payment disputes 63, 153
registering with 48–51
sending money 62–3
spending limits 53
using your account 52–3
verification process 53–8
plants, restrictions on
importing 92
postage scams 155–7, 167

postal orders 151
price comparison 37, 86,
  37–43
  gas and electricity 94–5
  insurance services 96
private health care 117, 120

returns policies 104
  eBay 108
routers, firewalls in 18

Sainsbury's website 84
saving data 24
saving money 9
scart sockets 90
screen readers 12, 15, 16
search engines 31–2
searching 31–6
  advanced 11
  choosing keywords 33–5
  eBay 147–151
  refining your search 34–5
secret questions 29
  eBay 136
security 11
  consumer protection 44–5
  credit card **46–7**, 63, 124
  of online payments 46–7
  updating 21
  see also anti-virus software;
    backing-up files;
    firewalls; fraud
seeds, restrictions on
  importing 92
self-drive holidays 124
shill bidding 155
shopping baskets 12

shopping online 8–9
  advantages 9, 66
  disadvantages 38, 40
shops, high street
  asking for cheaper deals 40
  visiting to view products
    38, 86
signing in and out 71
  eBay 140–2
size guides for clothing 103–4
sniping 173–4
software 12
  firewalls 19–21, 22–3, 25
spoof e-mails 12, **26–8**,
  153–4
spyware 12, **17**, 26
supermarket websites 70–1,
  82–4
  delivery 65
  express shopper 74
  favourites 73
  grocery pages 71–6
  offers and deals 76
  paying for shopping 80–1
  registering with 66–9
  shopping for groceries
    77–80

televisions 89–90
Tesco's website
  grocery shopping 66–81
text, resizing 14–5
travel
  booking flights 126–7,
    128–9
  disabled travellers 122, 128
  insurance 123

see also holidays

updating security 21
USB memory sticks 12, 24
user IDs 29, 134–5
Uswitch 94–5

VAT exemption 116
Virginsurfer 35–6

viruses 12, **17–23**
  anti-virus software 18,
  22–3

Waitrose's website 83
washing machines 86–7
web-friendly sites 35–6
websites 71–80, 82–4